Essential Guide to Medical Equipment Principles

This book addresses the needs of new clinical engineers, junior nurses, medics and operating department practitioners (ODPs) with regard to the fundamentals of many of the standard medical devices they will encounter in a hospital environment. The book can be used for both self-directed learning and also in a classroom environment as a textbook.

Essential Guide to Medical Equipment Principles introduces and provides the principles behind many of the common medical devices and equipment used within a modern hospital. Each of the seven chapters is designed to cover a particular type of medical device such as mechanical ventilators along with the physiology that must be understood in order to make sense of their application and engineering concepts. This fundamental knowledge will enable the reader to progress further in the years to come. The author uses diagrams throughout the book to allow the reader to get the most from the text and ensure that the most essential information is understood. It offers a comprehensive understanding of the physiological and engineering principles underlying medical equipment.

This book is intended for new technicians and engineers entering the clinical engineering field but can also be a valuable resource for a broader range of healthcare professionals, including operating department practitioners, neonatal unit practitioners, intensive care unit practitioners and other medical equipment users. With suppliers, manufacturers and in-hospital training, it will enable them to become safe and competent in the use and application of medical devices and equipment.

Essential Guide to Medical Equipment Principles

David Mulvey

CRC Press
Taylor & Francis Group
Boca Raton London New York

CRC Press is an imprint of the
Taylor & Francis Group, an **informa** business

Designed cover image: David Mulvey

First edition published 2025
by CRC Press
6000 Broken Sound Parkway NW, Suite 300, Boca Raton, FL 33487-2742

and by CRC Press
4 Park Square, Milton Park, Abingdon, Oxon, OX14 4RN

CRC Press is an imprint of Taylor & Francis Group, LLC

© 2025 David Mulvey

ISBN: 978-1-041-00355-7 (hbk)
ISBN: 978-1-041-00356-4 (pbk)
ISBN: 978-1-003-60941-4 (ebk)

DOI: 10.1201/9781003609414

Typeset in Times
by KnowledgeWorks Global Ltd.

*The book is dedicated to the memory of my closest friend,
Eric Nield (1946–2024), who sadly passed away shortly before the text of
this book was completed. I will always miss his friendship and council.*

Contents

Preface

I've had the privilege of knowing Dave Mulvey for over three decades. His mentorship in patient monitoring technology and principles early in my career profoundly influenced my path. Dave has dedicated his career to expanding his clinical engineering knowledge and sharing it with aspiring professionals.

Medical equipment is a cornerstone of modern healthcare, enabling better patient care, improved outcomes, and advancements in medical science. Clinical engineers play a pivotal role in ensuring the safe, effective, and efficient operation of this equipment. They inspect, maintain, troubleshoot, integrate, train, and manage costs associated with medical technology, ultimately contributing to improved patient outcomes and regulatory compliance.

This book was originally intended for new technicians and engineers entering the clinical engineering field but can also be a valuable resource for a broader range of healthcare professionals, including operating department practitioners, neonatal unit practitioners, intensive care unit practitioners, and other medical equipment users. It offers a comprehensive understanding of the physiological and engineering principles underlying medical equipment.

Operating theatres are particularly demanding environments, and this book is especially useful for engineers and support staff seeking careers in this field. The anesthetic machine is a critical component of any operating theatre, and its proper functioning is essential for patient safety. The book's sections on basic anesthesia principles and terminology are invaluable for those working in this area, including anesthetists, operating department practitioners, theatre nurses, clinical engineers, and technicians.

To excel in clinical engineering, professionals must possess a deep understanding of both the equipment's technical principles and the physiological principles underlying its use. Clinical engineering is not solely about medical devices; it's about how these devices interact with human patients. This book effectively bridges the gap between the physiological and electronic aspects of healthcare equipment, providing a comprehensive understanding.

In summary, this book equips clinical engineers and medical equipment users with additional knowledge and skills necessary to ensure safer, more effective and efficient operation of medical equipment, ultimately contributing to improved patient care and outcomes.

Prof. John Sandham
DProf CEng FIHEEM FIWBL MIET

About the Author

David Mulvey's medical engineering experience goes back over 40 years. Working in the commercial sector with S&W, Vickers Medical and his own company Artemis Medical Ltd, David held roles as Senior Engineer, Technical Training Manager and Technical Director. The final 10 years of David's career were spent in the UK's National Health Service as a Technical Training Manager and Senior Technical Support Engineer. For the 12 years David held the role of Technical Training Manager, he was able to develop and design training that had real relevance for clinical engineers, particularly those just embarking on their careers. David has run countless training courses and seminars not only in the United Kingdom but also in Europe. His training has gained a worthy reputation for a quality and practical approach to many complex clinical engineering subjects. Having been retired for the past 5 years, David has used this time to write this book to preserve and pass on his experience. David splits his time between Manchester, UK, and Orlando, FL, where he lives with his wife, Valerie.

Acknowledgments

The encouragement and support given to me by wife my Val and son Oliver have been crucial to the development of this book. Their proofreading has greatly improved the quality and readability of the text.

I thank my good friends Ted Mullen (NHS Scotland—retired) and Professor John Sandham (Chairman of EBME Expo, United Kingdom) for their invaluable support. They worked closely with me in the preparation of each chapter and offered valuable feedback and comments to ensure that the content is relevant for today's new clinical engineers, nurses, medics and ODPs. Thanks to Dr John Amoore who first suggested to me that I should write this book, more than 20 years ago.

I also want to thank to all those I worked with across my career of more than 40 years. A special thanks to Adrian Fisher my first manager in the medical equipment industry who, on three key occasions, gave me opportunities to learn and progress. I also thank Marc Gutierrez of Taylor & Frances Publishing for supporting me throughout the writing and agreeing to publish this book.

1 Heart and Circulation

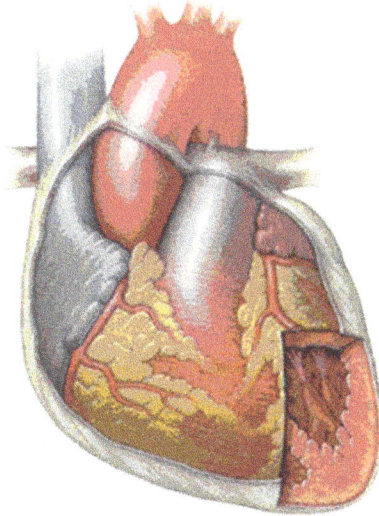

The five basic parameters on a patient monitor—electrocardiogram (ECG), non-invasive blood pressure (NIBP), temperature, SpO_2 and respiration—are either a direct measurement of heart functioning or are influenced by the heart's operation. As an engineer, we might approach the construction and functioning of the heart by describing it in four terms: first, its positioning and relevance as a pump; second, its construction; third, its mechanical functioning and fourth, its electrical control system. Adopting this method, clinical engineers should gain a clearer understanding of its importance.

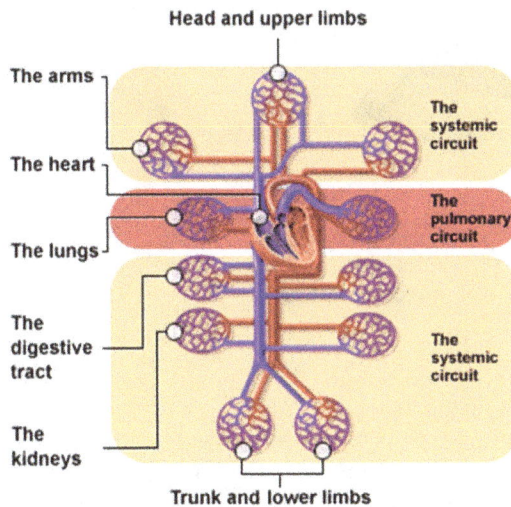

FIGURE 1.1 Basic circulation.

DOI: 10.1201/9781003609414-1

The simplified diagram in Figure 1.1 shows the heart is the center of the blood circulation system. Of course, the pump achieves the transportation of blood around the body. The colors used, red and blue, indicate that the blood at that point is either high in oxygen, red; or low in oxygen, blue.

The flow of blood through the lungs is known as the pulmonary circulation, and the lower circulation around the rest of the body (every other organ and tissue other than the lungs) is the systemic circulation. The change from high to low oxygen levels in the blood occurs in the systemic system due to metabolism. This is due to the burning or consumption of fats, carbohydrates and proteins within the various organs and tissues of the body in order to produce energy, growth and fight infection.

One way to consider this diagram is to think of it as the number 8. The upper circle sends blood from the right side of the heart to the lungs, and from the lungs it returns to the left side of the heart. From the left side it is then pumped out around the lower circle through the systemic system before returning to the right side of the heart.

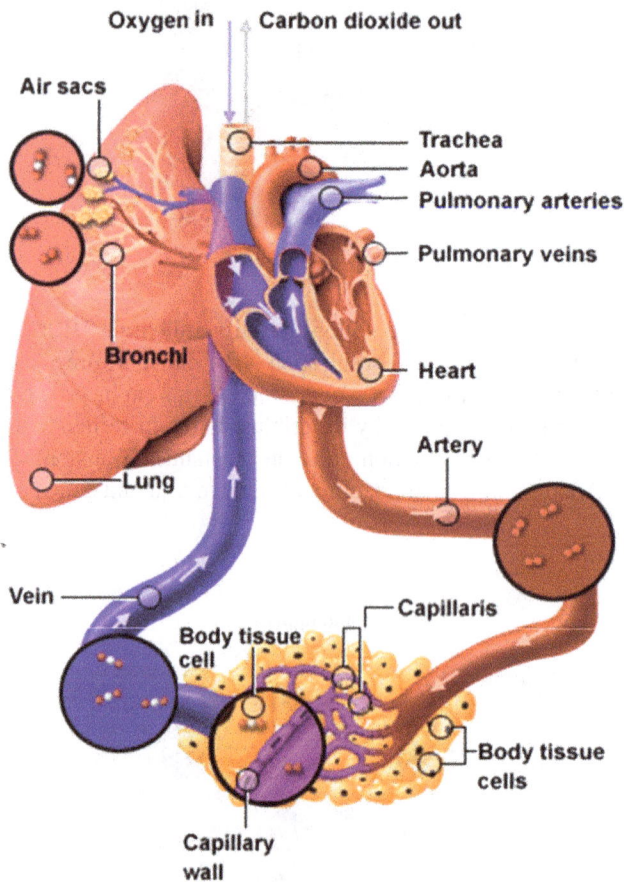

FIGURE 1.2   Circulation.

In studying Figure 1.2, you will note that there is oxygenated blood from the left side of the heart circulating through the aorta and the systemic system and returning as deoxygenated blood to the right side of the heart, from where it is pumped back into the lungs to be re-oxygenated and have the carbon dioxide (CO_2), which is a byproduct of metabolism removed.

As we breathe, we draw in air, which is rich in oxygen (approximately 21%) and has almost 0% carbon dioxide. The air inside the lungs then fills the minute air sacs within the lungs which are made up of even smaller cup-shaped cavities known as alveoli. The walls of the alveoli are so thin as to allow the gases oxygen and CO_2 to pass from the air sac into the bloodstream for oxygen, and from the bloodstream to the air within the alveoli for carbon dioxide. During the expiratory phase of a breath, CO_2 is then breathed out. The next inspiratory phase then brings in fresh gas with a high content of oxygen.

The left side of the heart receives deoxygenated blood (blue), then pumps it up through the pulmonary arteries to the lungs in order to circulate around and behind the alveoli sacks. Following gas exchange of oxygen and carbon dioxide, it then returns the blood to the right side of the heart through the two left chambers, which then go to pump it out through the aorta to the systemic system. Flowing through arteries, it then passes through capillaries. The blood finds its way to all the major organs and tissues of the body. Again, within the capillaries, the walls are extremely thin and allow gas exchange. This time, oxygen from the oxygenated blood is passed into the tissue of the organs and carbon dioxide is expelled from the organs into the returning blood stream. Carbon dioxide is the byproduct of metabolism, which is passed back into the capillary, and from there, the deoxygenated blood travels toward the veins, eventually returning the blood back to the right side of the heart.

One of the peculiarities of the circulation system is that the pulmonary veins carry oxygenated blood, and the pulmonary arteries carry deoxygenated blood. When talking about any other arteries or veins, we are always talking about the reverse, where oxygenated blood is carried by arteries and deoxygenated blood is carried by veins.

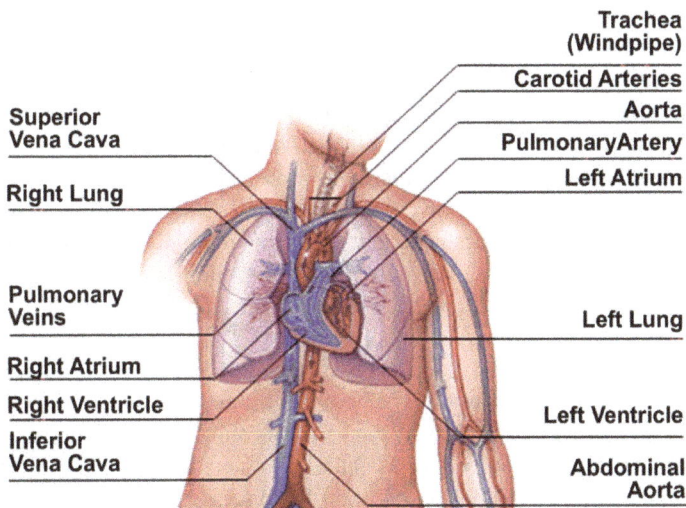

FIGURE 1.3 Major organs, arteries and veins.

The next step in understanding the heart and the circulation system is to study the major vessels that carry blood around the body. As you may have already gathered, these vessels divide into three categories. The first of these, which connect to the left side of the heart, provide oxygenated blood are the arteries. The flow of blood within the arteries is pulsatile in nature, meaning there is maximum flow during the heart's contraction phase (the systolic) and a minimum flow during its resting phase when the heart refills (the diastolic).

The second, which connects the arteries to veins, are the capillaries. These are minute pathways within the major organs and most tissues that allow little more than a single cell of blood to pass. It is here the nutrients and oxygen pass from the capillary into the surrounding tissues. And in the opposite direction, carbon dioxide, waste materials, and many other biological entities, such as hormones, are released back into the returning blood flow.

Finally, the third major component to the circulation system is the venous return. In essence, this carries the carbon dioxide back to the lungs and blood chemistry components. The blood flow in the venous return is at a lower pressure than the arterial blood flow, and it is smooth and constant in its flow. One other major use of the circulation system is to transfer heat from organs, such as the liver, around the body. When studying Figure 1.3, it is worth identifying the four major chambers of the heart—right atrium, the right ventricle, the left atrium, and the left ventricle.

The output from the left ventricle goes through the aorta (arch of aorta) and divides into the carotid arteries, and the abdominal aorta. These further divide throughout the body into such arteries as the gastric, brachial, and renal. Blood returning to the heart does so from two major veins—the superior vena cava and the inferior vena cava. The blood in these two major veins is the sum of blood taken from such veins as the iliac, renal and jugular veins.

<u>Physical Description</u>

- Mid-sternum (centre of chest)
- Size of a fist
- Apex (bottom) toward the left foot
- Four chambers, divided by the septum

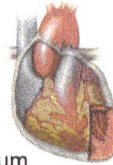

FIGURE 1.4   Cardiac anatomy

We will now study the construction of the heart itself (Figure 1.4). The heart is positioned mid-sternum, the center of the chest. It lies between the left and right lungs.

The second point above is that the heart is the approximate size of our fist. I would ask at this point that you take your right hand and make a fist without completely fully squeezing. When you look at the opening between your thumb and first finger and imagine the volume of fluid that might be able to flow between your fingers and your palm, you should see an opening of approximately 100 mL volume. If we were to say that the heartbeats at a rate of 60 bpm, then squeeze in your hand fully tight, you should be able to imagine injecting most of 100 mL. Therefore, you could produce an output of approximately 5–6 L/minute. This may not come as a surprise, but that is roughly the amount an adult heart can circulate around the body per minute.

The apex of the heart is at the very lowest point, and its angle points toward the left foot.

We now introduce the four chambers of the heart—the left atrium, the left ventricle, the right atrium and the right ventricle. A division between left and right is a wall of tissue known as a septum. Below is a collection of statements that should clearly explain the functions of the four chambers.

Left Side of the Heart	Right Side of the Heart
• Left atria and ventricle	• Right atria and ventricle
• Receives oxygenated blood from the lungs via the pulmonary veins.	• Receives deoxygenated blood from the superior and inferior vena cava.
• Pumps oxygenated blood to body (the systemic system).	• Pumps deoxygenated blood to lungs.
• Workload substantially greater than right side	• Workload less than left side
• Thicker walls than right side	• Thinner walls than left side

Layers of the Heart
- Epicardium
  ~ outer, white streaks of fatty tissue
- Myocardium
  ~ middle, muscle for pumping
- Endocardium
  ~ inner, fine lining, vessels connect

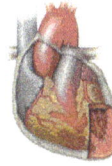

FIGURE 1.5 Cardiac anatomy.

If you were ever given the opportunity to witness the dissection a heart, you would find it is constructed of three distinct layers (Figure 1.5). The first outer layer is the epicardium, which appears as white streaky and fatty tissue and contains the heart's own blood vessels such as coronary arteries and veins. The second layer is the myocardium, which is the muscle layer and provides the contraction for pumping blood. It is within this layer that many cardiac problems originate. The third layer is the endocardium. This is a fine, smooth lining that the blood flow in each of the chambers passes across. When a clinician takes a stethoscope to listen to your heart, they are carefully assessing that the blood flow within the chambers is smooth and regular. A condition known as endocarditis can occur due to a viral infection. The endocardium layer may develop what might be considered wrinkles that cause turbulence to the blood flow that can be heard through a stethoscope.

Pericardium
- Sac surrounding heart
- Two layers
  - Outer layer tough tissue
  - Inner layer smooth moist lining
- Pericardial fluid decreases friction

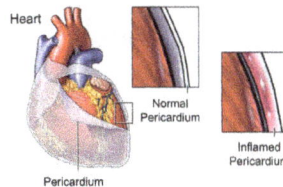

FIGURE 1.6 Cardiac anatomy.

As discussed earlier, the heart is located at mid-sternum between the lungs and the rib cage. In order to be able to constantly contract and refill, it is placed in a protective sac, the pericardium (Figure 1.6). This sac contains a pericardial fluid that enables the movement of the heart to occur without friction between it and the rib cage and lungs. Pericarditis is a condition that can occur when the outer layer of the pericardium becomes irritated and swollen. A patient with this condition will feel a sharp pain in the middle of the chest. This condition usually resolves itself without treatment in a few days.

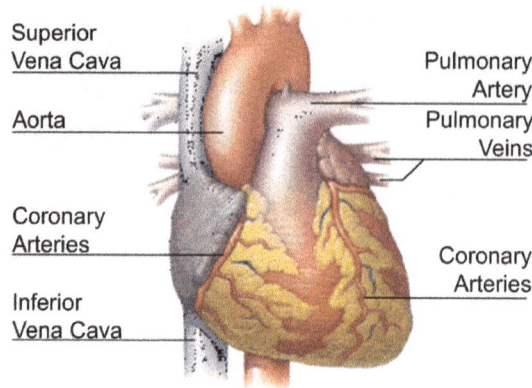

FIGURE 1.7 The external view of the heart.

As the heart is an organ, it requires its own blood supply. There are two main 'coronary' (relating to the heart) arteries, the right coronary artery (RCA) and the left main coronary artery (LMCA). Both arteries descend directly from an off shoot from the aorta (Figure 1.7). From these two coronary arteries, further arteries branch off to ensure adequate oxygenated blood reaches the whole of the myocardium muscle layer. As with all organs, there is a capillary bed within the myocardium and de-oxygenated blood from this is then returned via the coronary veins to the vena cava.

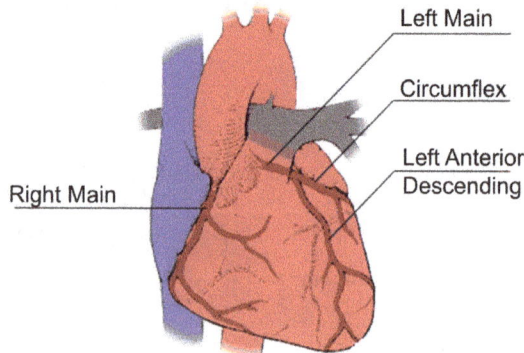

FIGURE 1.8 The coronary arteries.

As mentioned earlier, the two main arteries, left main and right main, divide further into circumflex coronary artery and left anterior descending artery, which then continue to divide into other smaller arteries (Figure 1.8).

Coronary arteries disease is a major medical problem today due to a combination of high-fat diets, smoking, genetic factors and environmental pollution. These factors can cause the build-up of fatty deposits within the coronary arteries. These fatty deposits, atheroma and the process of the build-up which is called atherosclerosis will eventually lead to the restriction of oxygenated blood flow and even a complete blockage. A restrictive flow in a coronary artery can produce sharp chest pains for the patient, a condition known as angina pectoris or angina, and the pain occurs from the myocardium not receiving sufficient oxygenated blood. If the severe restriction of the oxygenated blood flow through the coronary arteries is insufficient, then it may lead to what is commonly called a 'heart attack'. The correct term for such an event is a 'myocardial infarction' (MI).

The term infarction is a medical term commonly used to denote 'death', which is derived from a Latin term that translates to 'cram in to' or 'block'. If an MI takes place, then part of the myocardium muscle will die and cease to function. This will in many instances lead to the patient's death or severe heart injury.

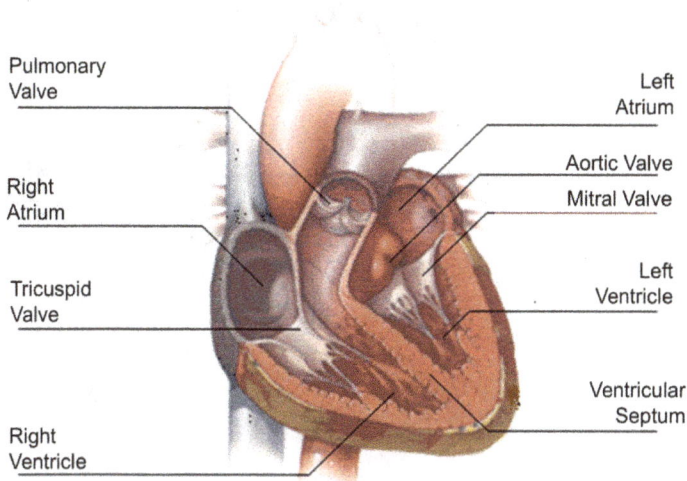

FIGURE 1.9   The internal view of the heart.

Before explaining the purpose and functioning of the heart valves, it is important to be aware of two terms that must be understood—'Systolic', it is when the heart is contracting in order to force blood flow, and 'Diastolic', when the heart relaxes enabling the atriums to refill. In order to ensure that blood flow around the body is only in a certain direction, there are four 'one directional' valves—two of these valves are internal within the heart and two external, positioned on the outputs from the heart (Figure 1.9).

Valves of the Heart

- *Semilunar:* Aortic valve and pulmonary valve
- *Atrioventricular (AV):* Tricuspid valve and mitral valve

Atrioventricular Valves (AV)—Internal

- Attached by chordae tendineae
- Prevents backflow during ventricular systole
- *Tricuspid*: Three leaflets, between the right atria and ventricle
- *Mitral*: Two leaflets, between the left atria and ventricle

Semilunar Valves (SV)—External

- Half-moon-shaped flaps
- Located in aorta and pulmonary arteries

When de-oxygenated blood returns from the superior and inferior vena cava, it passes into the right atrium which becomes full. At the outset of the compression phase (systolic phase) of the heart, the right atrium compresses and forces de-oxygenated blood through the tricuspid valve and into the right ventricle. At the same time on the left side of the heart, oxygenated blood that has returned from the lungs via the pulmonary veins enters the left ventricle and at the onset of the systolic phase forces oxygenated blood through the mitral valve into the left ventricle.

After the completion of the initial compression phase of the two atriums, a second more vigorous systolic phase occurs when both the left and right ventricle contract at the same time. During this part of the systolic phase, de-oxygenated blood is forced from the right ventricle through the pulmonary valve and into the pulmonary artery toward the lungs. Again, at the same time, the contraction in the left ventricle forces oxygenated blood through the aortic valve and into the aorta in order to circulate around the rest of the body (the systemic system).

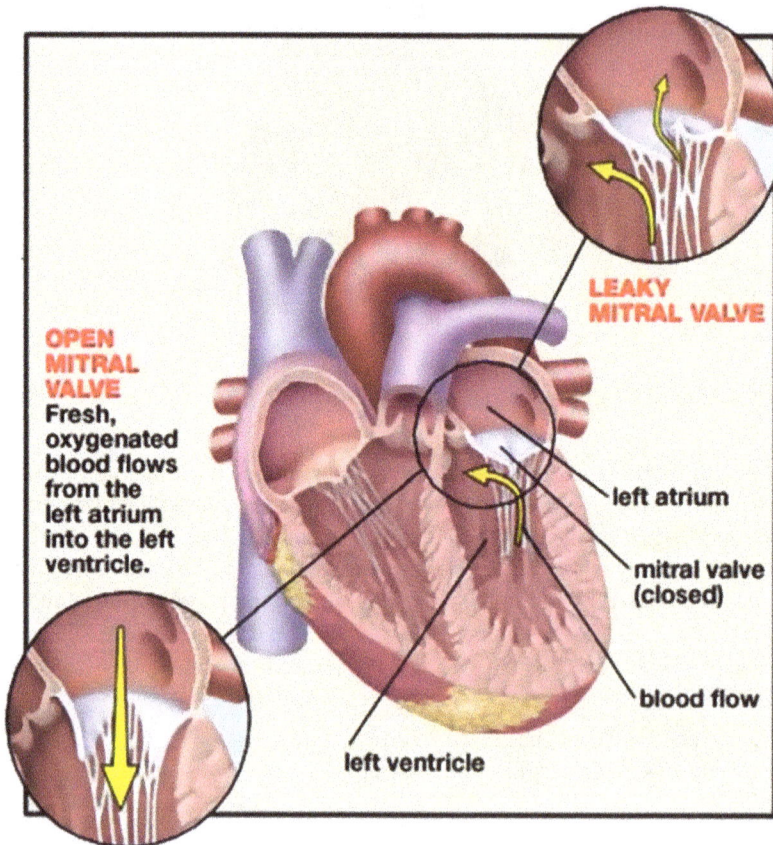

FIGURE 1.10   Valves and blood flow through the heart.

Figure 1.10 demonstrates the importance of the one-way valves such as the mitral valve in ensuring the blood flows in only one direction. Let's consider what will happen if the mitral valve allows blood to flow backwards during the systolic phase due to a leaky mitral valve. The first impact would be on the cardiac output volume that would drop as not all the oxygenated blood contained in the left ventricle is forced out through the aortic valve and around the systemic system. This would mean that organs and tissues such as the liver, brain and kidneys would receive an insufficient supply of oxygenated blood to meet their needs. The second consequence is that the low flow blood rate leads to an increase in the level of CO_2 in the blood stream that can alter the blood chemistry making it more acidic. This in turn adversely affects the functioning of many of the body's organs.

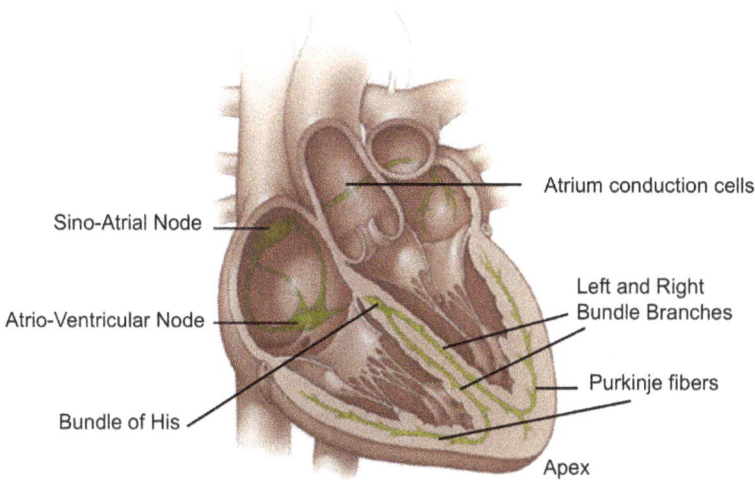

FIGURE 1.11 The conduction system of the heart.

What follows is a very simplified description of the heart's electrical control system (Figure 1.11), which is known as 'electrocardiograph' (ECG/EKG).

The fourth and final element to consider when understanding the heart is that it is controlled by an electrical control system. You may compare the heart with an electrical wiring diagram that you are already familiar with. The sino-atrial node (S-A node) can be thought of as setting the rhythm and timing for the sequence of contractions across the four chambers. If you have any musical understanding, you may wish to think of this as a 'metronome', setting the rhythm of the heart. An electrical impulse builds within the cells of the S-A node that eventually reach a critical point and is then transmitted along the atriums conduction cells that lie between the myocardial cells (muscle cells). This conduction of the electrical stimulus causes the atrium muscle cells to contract from their resting oval shape to a smaller circle shape, the effect of which is that the atriums contract forcing blood through the two valves, the mitral and the tricuspid valves and into the left and right ventricles.

Having the electrical stimulus pass over the two atriums, the electrical stimulus gathers at the atrio-ventricular node (A-V node). As the conduction of the electrical stimulus has been relatively rapid across the atriums, there is now a need to allow the mechanical contraction of the atriums to complete their full compression before allowing the electrical stimulus to travel further on to the ventricles. Within the A-V node, the electrical potential builds to an 'escape potential'. When the escape potential is reached, the electrical stimulus then passes rapidly down the 'Bundle of His', splits and continues down the 'left and right bundle branches'.

At the apex of the heart, the electrical stimulus then travels in an upward direction through the Purkinje fibers that are embedded within the myocardium around the left and right ventricles. This causes the ventricle myocardial cells to contract in the same way they did in the atriums, only this time the difference is that the compression starts at the Apex and moves up across both ventricles forcing blood up and out through the two external valves of the pulmonary valve and the aortic valve.

Following the systolic phase, the diastolic phase begins with the refilling of the two atriums and an electrical reset of the cardiac conduction electrical system. Also during this resting period, the myocardium muscle cells return to the resting oval shapes. In order to more understand the important function of the electrical control system, read a more detailed account of electrical control system in the chapter on ECG.

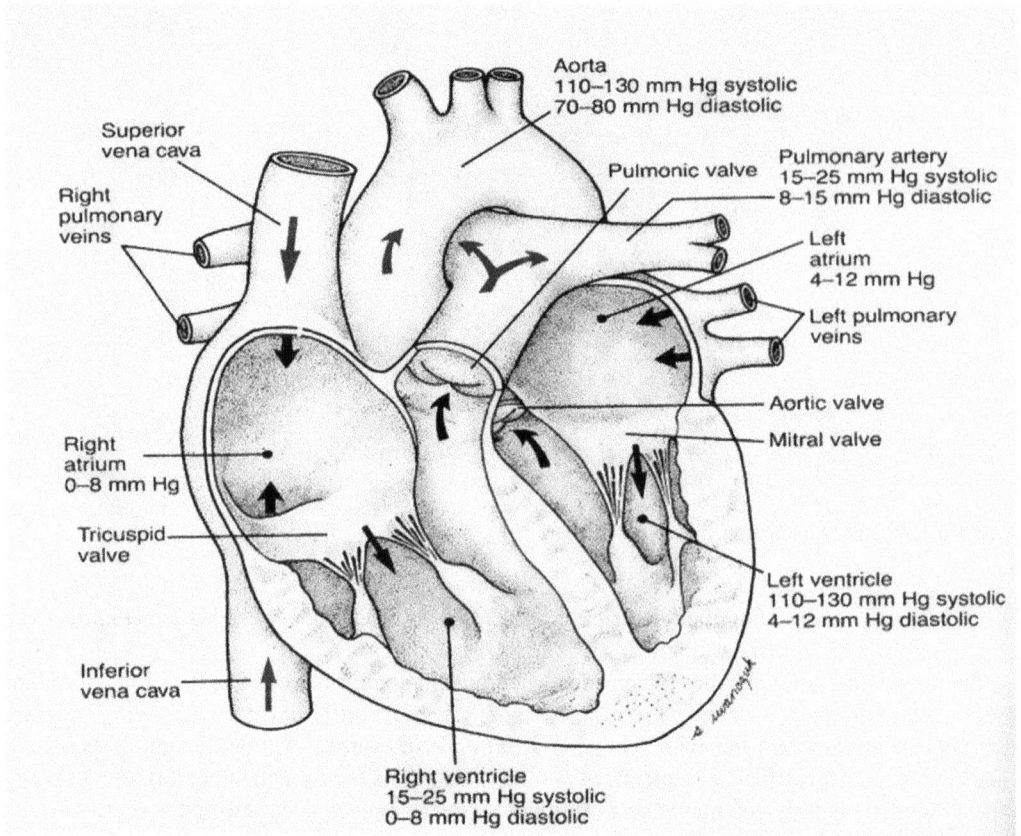

FIGURE 1.12 Pressures within the heart.

Figure 1.12 demonstrates the blood flow paths through the heart, and it will help to remember the entire process because it will be the basis for a more detailed understanding of the cardiac system. Below is a short list of statements that encompass all that has gone on before in this chapter.

Blood Flow through the Heart

- Atria relax (diastole)
- Right atrium fills from body
- Left atrium fills from lungs
- Pressure in atria exceeds ventricles AV valves (tricuspid and mitral) open
- Blood empties into relaxed ventricles (diastole)
- Atria contracts (systole) ~ fills ventricles
- Pressure equalizes atria and ventricles
- AV valves start to close
- Ventricles begin to contract (systole), pressure increases
- AV valves close completely
- Semilunar valves open
- Ventricles push blood into aorta and pulmonary artery
- Cycle is constant

2 Patient Monitoring

SECTION 2.1

ELECTROCARDIOGRAPH (ECG) MONITORING

This chapter on electrocardiogram (ECG) provides a detailed explanation of how the electrical conduction system works. It tackles the subject ECG to a fairly advanced level, and you may wonder if it is necessary. I firmly believe ECG is of great importance, it is worthy of the details it provides used in all emergencies areas such as operating rooms, emergency departments, cardiology departments and doctors' offices. A solid understanding of the subject of ECG will enable you to effectively relate it to the many problems that clinicians encounter in using ECG machines and monitors as you progress through your career as a clinical engineer.

Monitoring the electrical activity of the heart by ECG is a routine medical practice. The monitoring of cardio electrical activity is vital in many clinical situations such as the operating theater, cardiac care units, intensive care units and some medical and surgical wards.

ECG measurements (sometimes called EKG) are divided into two distinct methods—patient ECG monitoring (continuous three- or five-lead monitoring) using a video display to show the basic rhythm of the heart's electrical activity, and the more detailed analysis offered by a 12-lead ECG/EKG machine that is used intermittently by specialist Cardiac Technicians in order to gain a more profound understanding of electrical conduction across the heart.

Let's start with the term ECG.

- An electrocardiograph detects the electrical activity associated with the heartbeat and produces an ECG, a graphic record of the voltage versus time.
- Using electrodes placed on the skin, electrocardiographs record the small voltages (about 1 mV) that appear on the skin surface as a result of cardiac activity.

WHAT IS ECG?

The abbreviation ECG is formed from the three words.

- Electro-Cardio-Graphy
  - **Electro** (commonly relates to electrical activity)
  - **Cardio** (from the Greek word for heart)
  - **Graphy** (from the Greek word for writing)
- Electrocardio**GRAPHY** is the **METHOD** of studying electrical activity of the heart.
- Electrocardio**GRAPH** is the recording **INSTRUMENT**.
- Electrocardio**GRAM** is the **RECORDING**.

WHAT IS A CELL?

- *Electrical cell*: a single anode and cathode separated by electrolyte, used to produce voltage and current.
- *Biological cell*: the smallest unit that can live on its own and that makes up all living organisms and the tissues of the body.

Using the analogy of a battery cell you should then be able to relate to the idea that a biological cell shares one very important characteristic, that is, its ability to produce a small voltage.

DOI: 10.1201/9781003609414-2

As electrical/electronic engineers, we fully understand the important part battery cells play in today's world. The terminals of a single cell produce an electromotive force/potential difference (EMF) between the two terminals.

This characteristic is also to be found in the biological cell. Between the cell outer wall and the center of the cell there can be found a very small but significant charge of around 90 mV. This feature of these cardiac cells is central to the electrical signal (ECG) we find on the skin surface.

Before we move on to discuss the functions and characteristics of a cardiac cell, it is worth explaining in the simplest terms what a biological cell is. It's the smallest living organism consisting of three parts: the cell membrane, the nucleus, and, between the two, the cytoplasm. Within the cytoplasm lie intricate arrangements of fine fibers and hundreds or even thousands of miniscule but distinct structures called organelles. Cardiac cells have within the cytoplasm at rest a high concentration sodium (Na) and within the intra cellular fluid (the liquid between each cell) a high concentration of potassium (K). These two chemicals effectively give the cardiac cell its electrical property.

TRANSMEMBRANE ACTION POTENTIALS

In a resting state, cardiac cells maintain a negative (−) charge on the inside of the cell membrane and a positive (+) charge on the outside of the cell membrane (Figure 2.1.1).

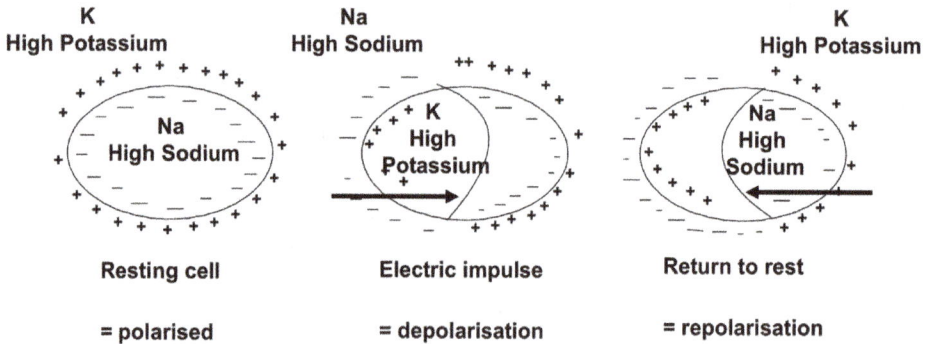

FIGURE 2.1.1   Transmembrane Action Potentials.

Electrical stimulation causes Na⁺ ions to move across the cell membrane to the inside so that the cell becomes positive on the inside. During repolarization, the cell is restored to the resting state.

Figure 2.1.2 addresses the ability of the cardiac cells to change polarity. Used here are analogies that electronic engineers should already be aware of.

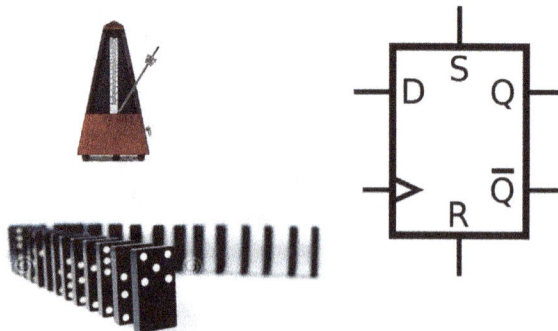

FIGURE 2.1.2   Cardiac cells principal.

A metronome is a bi-stable device that has two equal states, and it alternates between the two states at a given rate of change (continuous flip/flop). The cells within the atrial ventricular (AV) node in the heart should be thought as the 'heart's pacemaker' and initiates each heartbeat.

A line of dominos is a mono-stable device that remains upright until stimulation causes the first to fall causing a 'domino effect' along the line of dominos. Each domino acts as a stimulation to the next. Given a set period of time, they reset to the upright position. This is very much how the conduction cells of the heart behave.

Figure 2.1.3 should be committed to memory as it is the basic wiring diagram of the electrical signal path across the heart. Memorizing this will enable you to picture where physically a conduction problem might lie.

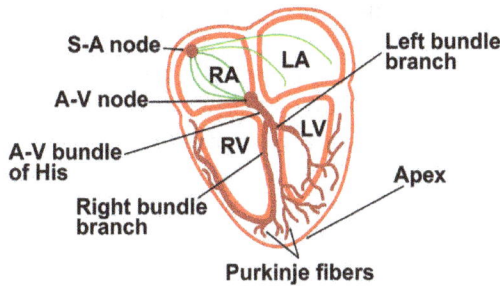

FIGURE 2.1.3   Conduction system of heart.

The conduction cells lie within the heart muscle (myocardium) and as they change state, they provide a stimulus not only to the next cardiac conduction cell but also to the adjoining oval muscle cells that then contract. The overall effect is at the heart's chambers—the atriums and the ventricles—that pump blood. *Please continually refer back to this diagram as you proceed through the detailed description of the ECG signal and its path across the heart.*

ECG WAVEFORM—OVERVIEW

The ECG waveform appears at first sight to be complex. It is in actual fact just the representation of the electrical stimulus path across the heart.

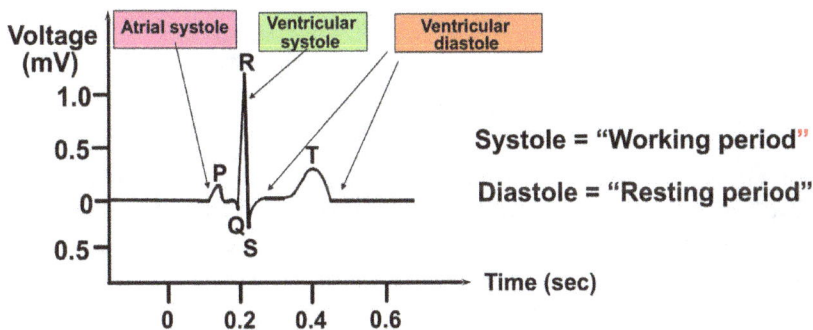

FIGURE 2.1.4   Normal ECG in relation to the mechanical activity of the heart.

Electrocardiogram is the recording of electrical activity of the heart (Figure 2.1.4).

THE NORMAL ELECTROCARDIOGRAM (SINUS RHYTHM)

ECG reflects the electrical events generated by the heart muscle.

The P-wave originates in the S-A node, the sino-atrial node which is the pacemaker of the heart. This is a collection of bistable cells that work together to provide a steady continuous regular change in 'polarization'. The **P-wave** corresponds to the rapid activation of the atria resulting in atrial systole (atrial contraction), during which blood is forced down through two internal valves. The tricuspid valve between the right atrium and right ventricle has returning venous blood that is pumped to the lungs (pulmonary). The mitral valve sits between the left atrium and left ventricle and receives oxygenated arterial blood from the lungs and the forces it out around the rest of the body (the systemic circulation).

The ECG signal, having passed over the two atriums, arrives at the atrioventricular node (AV). Here, a small delay is introduced to allow the two atriums to complete their contraction of pumping blood into the two ventricles. As shown in Figure 2.1.4, the QRS complex, which will be explained later in this chapter, corresponds to the rapid spread of the electrical activation through the ventricles resulting in ventricular systole or 'working period'. The left ventricular wall is much thicker than the right one and has thus greater influence on the ECG signal.

The **T-wave** corresponds to the ventricular diastole or 'resting period' when blood fills the ventricles before the next 'working period'. It is during this period that the cardiac conduction cells reset, repolarization.

THE ECG WAVEFORM—DETAIL

We now approach the point where we begin to relate the ECG waveform to the transmission of the electrical signal across the heart and the heart's mechanical response to it.

Being able to pick up the heart's electrical activity is fundamental. Figure 2.1.5 shows the conventional electrode placements used to do this.

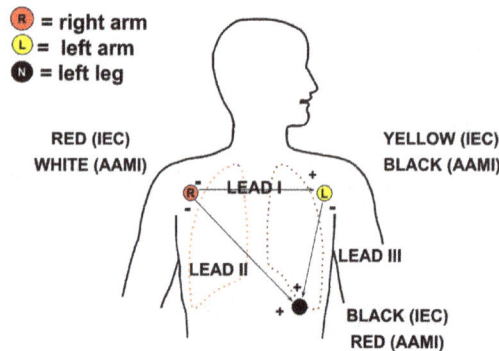

FIGURE 2.1.5   Einhoven's Triangle. 3-lead ECG.

This figure has been positioned at this point in order to demonstrate that a 'positive deflection' on the ECG waveform is due to the direction of the signal in Lead II toward the left leg. This is true for each 'lead's position' with regard to the 'positive electrode' for each lead. Therefore, a 'negative deflection' represents a signal moving away from the positive electrode. A more detailed explanation will be given further on in this chapter.

Figures 2.1.6–2.1.16 go into great detail in explaining each wave, interval and complex within the ECG signal. What is offered here is the level of insight needed to use the ECG signal as a diagnostic tool. You may find it useful to look at the time durations given in terms of milliseconds (ms) rather than seconds (s). Therefore, 0.11 seconds is 110 milliseconds. A 60 bpm (beats per minute) ECG rhythm will mean that one complete contraction cycle is 1 second long or 1000 milliseconds.

P Wave

- Pacemaker site = SA Node
- Duration: 0.11 seconds or three small boxes (or less)
- Shape: Rounded and smooth
- Amplitude of 0.5 to 2.5 mm in lead II

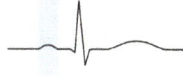

FIGURE 2.1.6

Originating in the sino atrial node (SA Node/Pacemaker) this bistable signal sets off a transmission of electrical stimulation to the conduction cells across the atrium heart chambers causing a domino effect (mono-stable cells). This stimulation produces a contraction of the myocardium muscle cells within the atriums as it passes. The two upper atrium chambers contractions forces blood down through the mitral and tricuspid valves into the two ventricle chambers.

PR Interval

- Onset and end: Starts at end of P wave, ends with start of QRS
- Duration :0.12 to 0.20 seconds, HR/ age dependent
- Significance: AV node/bundle of HIS pathway functions

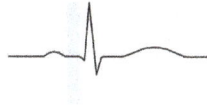

FIGURE 2.1.7

As the mechanical response to the electrical transmission is slower than the conduction of the electrical stimulation across the atriums, a slight delay is required before the conduction of the electrical stimulation to the ventricles. This delay is achieved within the AV node. You might think of this as a capacitive/resistance device that has to reach a certain voltage potential before allowing the forward transmission of the signal to the ventricle chambers.

Q Wave

- First downward deflection in the QRS complex
- Represents septal depolarization
- Duration: normally less then 0.01 seconds;
- Significance: Abnormal Q (1) greater than 0.04 seconds wide or, (2) greater than 25% of R wave amplitude

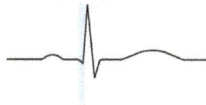

FIGURE 2.1.8

Once the signal leaves the AV node, it does so from the top of the node, thus causing a slight 'negative deflection'. It then turns around the AV node and heads south toward the AV bundle of His. This is followed by an almost vertical positive deflection to 'R'. This is the 'depolarization' of the septum as the signal heads toward the 'apex'.

S Wave

- Is the first downward deflection after an R wave

- Duration: Usually 0 to 0.05 seconds

- Amplitude: .8 mv

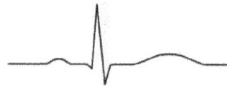

FIGURE 2.1.9

From the apex of the heart, the signal then travels from the apex along the Purkinje fibers, heading back across the two atriums, both anterior and posterior (front and back of the heart). As the signal is now moving away from the left leg towards the positive electrode in Lead II, we now see a sharp negative deflection. This represents depolarization and contraction of both ventricles.

QRS Complex

- Combination of Q, R, and S waves

- Represents ventricular depolarization, atrial repolarization

- Largest wave of ECG

- Pacemaker site: Normally SA node, however, ectopic or escape pacemaker in atria or AV junction

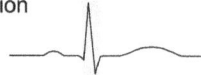

FIGURE 2.1.10

As you may now have realized the many elements of an ECG waveform are given terms such as 'complex', 'wave', 'interval' and 'segment'. This enables us to distinguish each period of the ECG into electrical activity that represents a biological and mechanical action. The 'QRS complex' is one such period representing the depolarization of the septum and ventricles and the onset of the contraction from the apex up across both ventricles.

THE QRS COMPLEX

- *Onset*: where the first wave of the complex begins to deviate from the baseline.
- *Ends*: where the last wave flattens out, at or above the baseline.

- *Duration*: 0.10 seconds or less in adults, or 0.08 seconds or less in children.
- *Amplitude*:
  - No less than 5 mm in Leads II, III, avF and V1–V6;
  - No greater than 25–30 mm in V1–V6.

The vertical position of the 'QRS complex' will vary greatly when looked at through other leads such as V1–V6 as their positive electrode is not in the same vertical and horizontal position relative to the heart and the signal path. A good example of this is Lead V1 that has its positive electrode placed on the left side of the chest level both horizontally and vertically with the mid-septum. Thus, we see a QRS complex that is equal in its positive and negative deflection.

The Absolute Refractory Period

- Shape: Predominately positive (upright), negative (inverted) or partly negative and partly positive
- Significance: Electrical impulse has reached Purkinje fibers without delay. Normal depolarization of ventricles occurred.
- **Note:** The Absolute Refractory period is represented from the R wave to peak of T wave

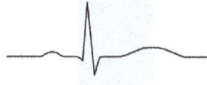

FIGURE 2.1.11

It's during this period that depolarization has taken place in the ventricles, the conduction cells in the septum (AV bundle of His, left and right bundle branches) and Purkinje fibers will no longer react to any further stimulation until they have almost all repolarized (reset). This is important as it is in the refractory period that the two ventricles do not attempt to 'pump' again as they have just emptied.

ST Segment

- Absolute Refractory Period (ARP) of ventricles
- Onset : End of QRS complex
- End: Beginning of T wave.
- Junction between QRS complex and ST segment called "**J point**" /"**J Junction**".
- Duration: 0.20 second or less, HR dependent

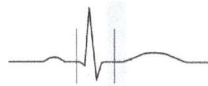

FIGURE 2.1.12

The significance of the ST segment is return of the signal to the 'base line'. Two points are marked— first, the isoelectric line just before the 'Q' wave (the base line), and second, the 'J point' that is selected between the 'S' wave and the 'T' wave.

ST Segment

- Amplitude: Normally flat (isoelectric), can be elevated or depressed less than 1.0 mm at 0.08 second (2 small squares) after J point.

- Significance: Normal ST segment followed by normal T wave shows normal repolarization of both ventricles

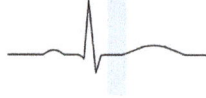

FIGURE 2.1.13

The deviation in height between these two points allows for the interpretation of a number of cardiac conditions including previous heart attacks, ischemia and the effect of an anesthetic agent on the heart.

T Wave

- Ventricular repolarization

- Onset : The first deviation from ST segment. If no ST segment, begins at end of the QRS.

- End: Returning to baseline considered the end.

- Duration: Usually 0.10 to 0.25 seconds

- Amplitude: Less than 5 mm

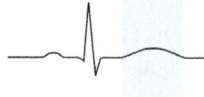

FIGURE 2.1.14

This repolarization of the ventricles plays an important part in the diagnosis of cardiac damage, the speed and amplitude of this wave and the return to the isoelectric base line.

T Wave

- **Shape:** Sharply rounded slightly asymmetrical
- **Significance:** Normal repolarization of right and left ventricles
- **Note:** The Relative Refractory Period is from peak to end of T wave.

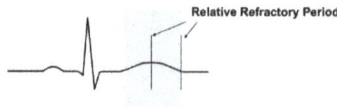

FIGURE 2.1.15

The relative refractory period is when it may be possible to stimulate the conduction cells within the ventricles, producing an 'ectopic beat' and thus a pumping action from the ventricles that produces little flow.

QT Interval

- Age, sex, HR dependent
- Onset: Beginning of QRS
- End: Beginning of T wave.
- Duration: Generally 0.32 to 0.40 seconds.
- Significance: Depolarization and repolarization of ventricles and time of systole in cardiac cycle.

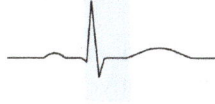

FIGURE 2.1.16

It is this 'interval' that shows the most variation in its duration with a change in heart rate (HR). The slower the HR, the longer the duration; the faster the HR, the shorter the QT period. Children with faster HR have a shorter 'interval' of 0.32 seconds. For adults with 'bradycardia', a longer 'interval' of more than 0.40 seconds is to be expected.

ECG MONITORING

LEADS AND CONFIGURATION

The next short series of figures demonstrates how we 'sense' the ECG signal using electrodes placed on the skin. The position and signal direction are fundamental to how we see the ECG signal on both 'monitors' and 'ECG machines'.

The precision placement of 'wires' on the chest is critical to being able to obtain a good reliable ECG. Only in using the positions shown (Figure 2.1.17) does an ECG signal make any sense.

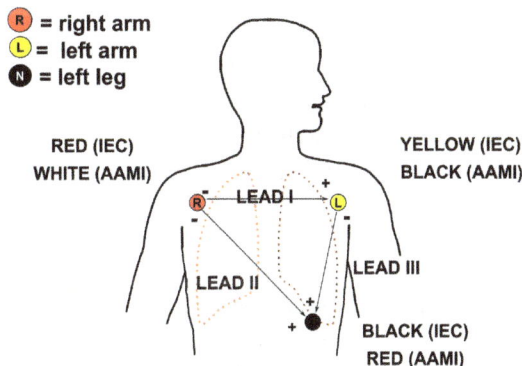

FIGURE 2.1.17   Lead ECG.

EINTHOVEN'S TRIANGLE

The term 'leads' is somewhat confusing as it refers to the signal measurement between two 'points' (wires/cables). Leads refer to a precise placement on the chest not an individual wire/cable.

THREE LEAD ECG (LEADS I, II AND III)

The three-lead system is commonly used in routine monitoring. The Einthoven's anatomic electrical triangle provides the basis for placing the electrodes in relation to the normal direction of the heart's electrical activity.

Electrodes for the standard leads are placed on the right and left arms and on the left leg. The voltage variation is measured from negative to positive electrodes, the third electrode being the 'Reference' grounding electrode to reduce the electrical interference. The standard lead selections are:

Lead I between right arm (−) and left arm (+). Left leg (reference)
Lead II between right arm (−) and left leg (+). Left arm (reference)
Lead III between left arm (−) and left leg (+). Right arm (reference)

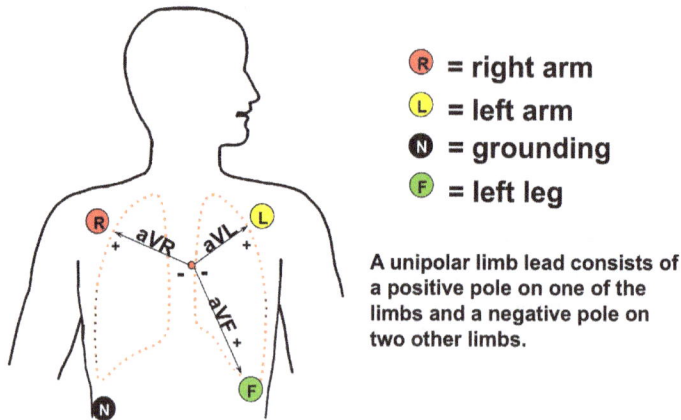

R = right arm
L = left arm
N = grounding
F = left leg

A unipolar limb lead consists of a positive pole on one of the limbs and a negative pole on two other limbs.

FIGURE 2.1.18   Augmented limb leads (aVR, aVL, aVF).

'Augmented Leads' are an option of creating a further three-lead positions from the same placements of the 'Three Wires' that created Leads I, II and III (Figure 2.1.18). Internally within the patient monitor or ECG machine connecting the negative and reference wires together, we create a middle electrical position between the two limb wires. Thus, surprisingly we are able to produce some six EGC leads.

Thus:

- Lead aVR has the right arm positive (+) and the left arm and the left leg connected together become the negative (−).
- Lead aVL has the left arm positive (+) and the right arm and the left leg connected together become the negative (−).
- Lead aVF has the left leg positive (+) and right arm and left arm connected together become the negative (−).

A more complex form of ECG monitoring is afforded by the use of 5- and 12-lead ECG monitoring. The conventional Einthoven's triangle gives what might be thought of as a 'vertical' view of the heart, whereas a 12-lead configuration also adds a 'horizontal' view.

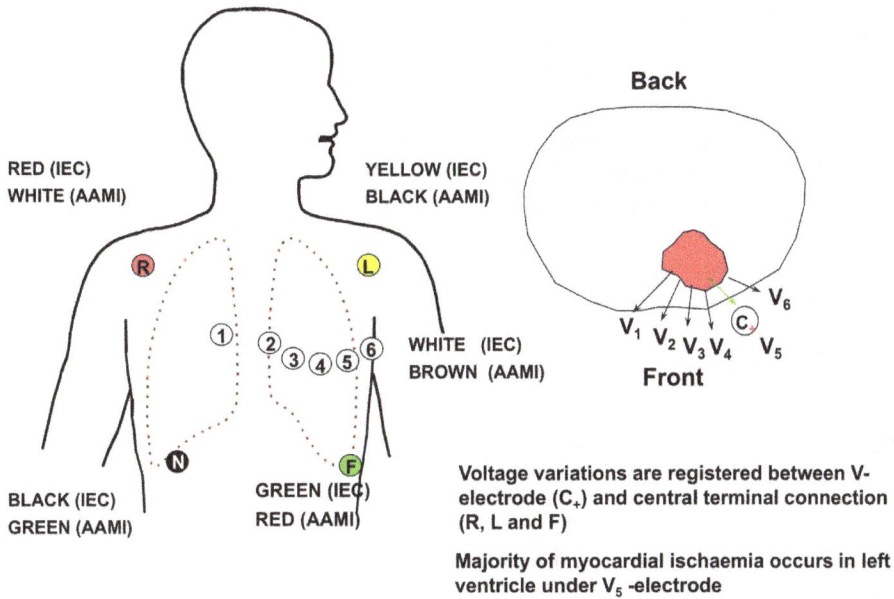

FIGURE 2.1.19   5 and 12-Lead ECG.

The use of 5- and 12-lead ECG signal analysis offers a greater degree of ECG interpretation (Figure 2.1.19). Using five leads, we enable gathering of four positions from the 'Limb' leads plus a fifth when we position at 'V5', that is, close to the 'apex' of the heart and 'left ventricle'. It is within this area that we can often see the effect of 'myocardial ischemia' (poor perfusion) or evidence of previous 'heart attacks' (myocardial infarction.). The additional 'V leads' enable greater clarity of the electrical conduction particularly across the two ventricles.

The following sections address the many practical issues in the use of ECG monitors and machines.

FIGURE 2.1.20   Clinical objectives of ECG monitoring.

Monitoring as opposed to the use of a 12-lead machine is continuous and has limited aims (Figure 2.1.20). HR is determined by the counting of the 'R waves' intervals over 1 minute. The 'basic ECG wave-forms' when we look for significant deviation from a normal ECG such as 'ectopic beats' and significant changes in 'ST-segment' may indicate problems such as previous heart attacks or ischemia.

ARTIFACTS AND INTERFERENCE: SOURCES OF PERIOPERATIVE ECG

Artifacts and interference can arise from a number of sources, chiefly:

1. Electrosurgery interference (150 Hz –>)
2. Equipment using alternating current (50–60 Hz)

3. Low frequency noise (0.1–100 Hz)
4. Defibrillation
5. Muscle noise (electromyograph—EMG)

There are many and varied difficulties for obtaining a clear and reliable ECG trace. The small ECG signal can often be overwhelmed by any of the issues above. Chest electrodes should be kept away from a high-power diathermy scalpel by placing the electrodes on the patient's back during chest surgery. Ensuring that the ECG trunk cable is dressed away from any mains cable will reduce 'mains pickup'. Moving the left and right electrodes further up toward the scapula (the collar bones) means that the low-frequency breathing variation of the chest impedance, that can cause the baseline of ECG to 'wander', is eliminated. Ensuring, where possible, that the patient has minimal 'muscle tremor' will also improve the clarity of the ECG signal.

Minimizing Interference

Interference can come from a number of sources. Where possible, minimize interference from:

- other electrical equipment (e.g., infusion pumps and blankets)
- unclean and unshaved skin
- poor quality electrodes (e.g., from drying out)
- dirty leads
- broken lead wires (e.g., those with no signal)
- misconnection

The correct application of ECG electrodes is paramount in avoiding poor signals. The skin should be shaved if necessary and then cleaned with an alcohol wipe. As the electrode is attached to the skin, it should be 'rolled across' the skin ensuring that air is not trapped between the electrode gel and the skin. Always ensure that the electrodes have been stored correctly, within the expiry date if given and do not appear to be dried-out. The ECG cable should be inspected for damage and that the cable clips are clean. Some cables systems come in two parts with a trunk cable that the individual limb and precordial leads plug into. Often these may be missing or placed in the wrong order. *Be aware that there are two systems of cable coloring codes: European (IEC) and North American (AAMI).*

Filtering ECG

When monitoring, the following filter settings can be applied for the various clinical assessments:

- *Monitor bandwidth*: 0.5 z to 40 Hz
- *ST bandwidth*: 0.05 Hz to 40 Hz
- *Diagnostic bandwidth*: 0.05 Hz to 100 Hz

Selectable filtering is often available on modern patient monitors in order to allow the clinician to open the bandwidth to allow more subtle changes to be observed in the ECG waveform. The diagnostic mode is used in such areas as 'Cardiac Heart Care Wards' when a 12-lead cable is used for continuous or intermittent monitoring. The 'ST mode' is particularly useful in the operating room for the anesthetist who will often select 'ST' mode in order to see changes in the 'ST segment' due to the use of anesthetics and the cardiac effect on the patient.

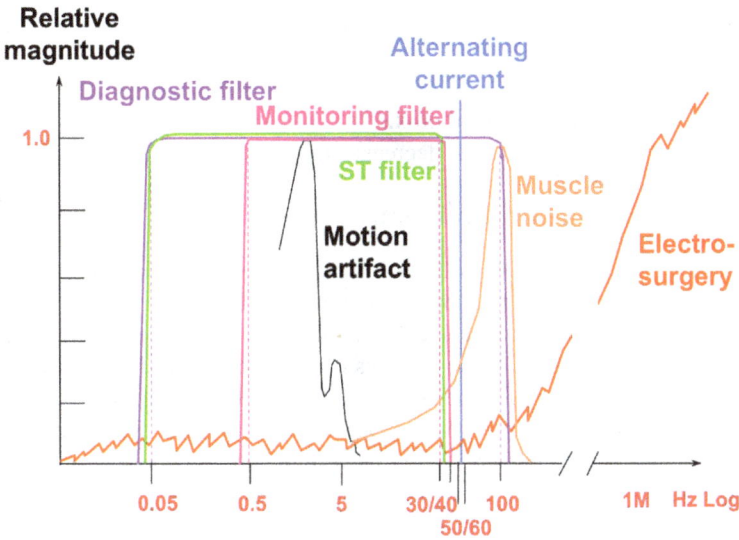

FIGURE 2.1.21 ECG filtering.

In Figure 2.1.21 we are able to see the level and frequency range of some of the sources of electrical and motion interference that can degrade the ECG signal, muscle tremors, respiration and electro-surgery being the main culprits.

IN CONCLUSION

You should now be aware of the importance of ECG measurement as a diagnostic tool in the field of cardiology. Almost all patients in operating rooms, intensive care and many other medical settings are monitored for ECG because cardiac disease and complications are a major source of serious illness and morbidity. There are many potential pitfalls to obtaining a clear ECG, but as outlined in this chapter, there are practical steps you can take to avoid these.

SECTION 2.2

IMPEDANCE RESPIRATION

Breathing is vital for life. You won't survive very long without it—minutes at most. Monitoring a patient's breathing is therefore of critical importance.

Respiration monitoring can take place in two distinct situations. The first is in the spontaneously breathing patient, whether conscious or unconscious. The second is the patient who is attached to a mechanical ventilator. This chapter addresses the most common form of respiratory monitoring for the spontaneously breathing patient.

As electronic engineers, you may be aware of a radio transmission system called amplitude modulation (AM). This is somewhat similar to the electronic technique we use to produce the respiration wave form and rate measurement on a patient monitor. An AM signal has a carrier frequency whose amplitude is varied with the sound signal that it carries. When this arrives at the receiver, it is demodulated by the removal of the carrier frequency signal, which leaves us with the far slower AM element which then can be amplified and turned into sound.

THORACIC IMPEDANCE

What is thoracic impedance?

- It is the electrical resistance of the thorax.
- It increases during inspiration and decreases during expiration, since air has a higher impedance than tissues.
- Impedance change is interpreted as a breath.

As the chest expands and contracts with each breath, its electrical impedance (resistance) also changes. As air has a high electrical resistance, it therefore follows that during the inspiration phase when the chest is full of air, the resistance across it will increase. Consequently, during the expiration phase, its resistance decreases. These changes can be seen as modulating a low voltage, high frequency carrier signal applied across the chest via the chest ECG electrodes.

This means that the ECG electrodes are not only capable of monitoring the 1 mV ECG signal (input signal from chest leads), but also applying a high-frequency respiration carrier signal of approximately 100 µV at 10 kHz (output signal to chest leads).

MEASURING IMPEDANCE RESPIRATION

Using ECG leads, it is possible to obtain two vital signs parameters from just one set of connections. This reduces the number of connections that need to be made to the patient.

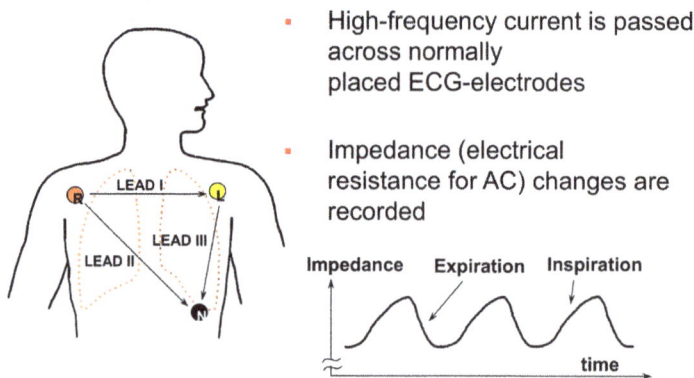

FIGURE 2.2.1   How to measure impedance respiration.

The usual lead selection for impedance respiration monitoring puts Lead 1 between the right arm and the left arm. This position often offers the greatest degree of respiratory impedance change during respiration (Figure 2.2.1) The electrodes are sometimes moved lower on the chest and even around to the side in order to maximize the respiratory change. Lead 2 is often used for respiratory monitoring of neonates as they use their tummy muscles to supplement the diaphragm efforts. Be aware that the amplitude of the signal will vary greatly between individuals and in itself it is not relevant, but over time on an individual may vary and give cause for concern about the depth of their breathing.

APPLICATIONS OF IMPEDANCE RESPIRATION

Impedance respiration is used for:

- monitoring the patient's respiration rate
- detection of apnea
- it is used in SCBU, recovery rooms, day surgery, patient transport and ICU.

The rate and depth of an individual's breathing is a very good indicator of their respiratory function. It is closely monitored following anesthesia when the recovering patient is now beginning to breathe for themselves spontaneously. Clinical staff will watch carefully for a patient in the recovery area who may suffer from respiratory collapse once removed from the ventilator. Apnea, which is the total cessation of breathing and not to be confused with a lower respiratory rate, can often occur. In neonatology, this is a very real problem for neonates who are very susceptible to apnea attacks. The more sophisticated monitors are able to count the number of respiratory apneas and their frequency.

FACTORS AFFECTING THORACIC IMPEDANCE MEASUREMENT

Thoracic impedance measurement can be affected by:

- Deep or superficial breathing
- Electrode positioning and adhesion to the skin
- Motion
- Cardiac artifact

Thoracic impedance measurement does not detect obstructive apnea.

Often obtaining a good respiratory impedance signal can be difficult. Individuals who are smaller and sometimes older breath less deeply than larger individuals. This means that the positioning of electrodes may need to be changed in order to maximize the respiratory signal. Vibrations caused by motion, such as a trolley's or travelling in an ambulance, can also negatively impact the impedance respiratory signal.

The heart, which is placed between the left and right lungs, causes a small change in the respiratory signal as it fills and empties. This can be eliminated by moving the electrodes slightly further up the chest when using ECG Lead 1.

Crucially, the monitoring of respiratory impedance does not detect obstructive apnea. The chest may appear to rise and fall, but the amount of air is severely limited due to the obstruction. It is in this situation that it is not a reliable indicator of sufficient ventilation.

CONVENTIONAL IMPEDANCE RESPIRATORY MEASUREMENT

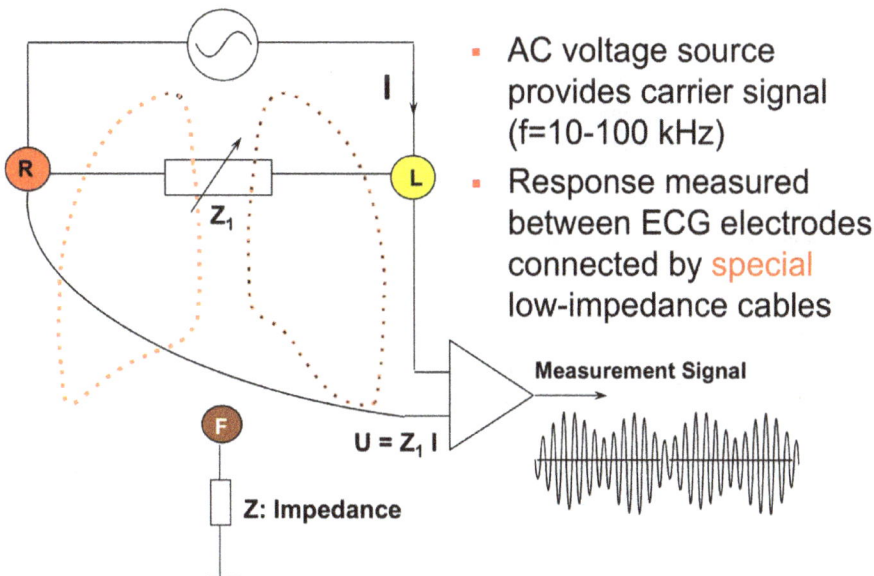

- AC voltage source provides carrier signal (f=10-100 kHz)
- Response measured between ECG electrodes connected by special low-impedance cables

$U = Z_1 I$

Z: Impedance

Measurement Signal

FIGURE 2.2.2   Conventional impedance respiratory measurement.

Figure 2.2.2 depicts the electronic principle behind impedance respiration. The circle with the sinusoidal waveform is the AC generator that produces a high frequency signal between 10 kHz and 100 kHz. This is applied to the right-arm electrode and the left-arm electrode, and the current flows across and through the chest as the impedance Z1 varies between inspiratory and expiratory, so the amplitude of the signal applied to the differential amplifier (triangle) will vary. Once within the monitor, the signal is then passed to a demodulator that removes the high-frequency signal and leaves only the variation in the amplitude due to breathing. The software within the monitor is able to interpret the amplitude change and count as a rate and display as a waveform.

SECTION 2.3

NONINVASIVE BLOOD PRESSURE (NIBP) MONITORING

Noninvasive blood pressure (NIBP) monitoring is a readily available tool for clinicians to understand the condition of a patient's cardiovascular system. Used routinely in all types of clinical situations, it offers a clinician a rapid and relatively accurate picture of the patient's blood pressure.

Critically, an NIBP offers a relative not an absolute reading. This means that it is lacking in absolute accuracy compared with an invasive blood pressure (IBP) monitoring system. There are a number of reasons why NIBP readings are not as accurate as an IBP reading, not least that it is a snapshot of the patient's blood pressure at any given moment.

SYSTOLIC AND DIASTOLIC PRESSURES

- *Systolic pressure* is the working phase of the left ventricle, in which it contracts and pushes blood into arteries.
- *Diastolic pressure* is the heart's resting phase, in which it fills with blood.
- *Mean pressure* is the average pressure in the cardiovascular system.

Systolic and diastolic are the two readings generally quoted from taking an NIBP reading. The systolic representing the contraction phase of the heart shows the maximum pressure in the system during the cardiac cycle. Diastolic is pressure measured during the refilling of the heart, the resting phase. You may have assumed that the systolic pressure is the one that the clinician will be most concerned with, but in actual fact it is the diastolic pressure that is often a better indicator of a patient's cardiac condition. New research however is suggesting that systolic pressures are also important indicators of hypertension.

The systolic pressure will often vary greatly with exercise and stress, whereas the diastolic pressure, showing the resting pressure, tends to indicate possible underlying cardiovascular medical conditions.

The mean pressure is useful for determining the general trend of the blood pressure as a whole and can be monitored over time to ensure stability.

COMMON METHODS OF NIPB MEASUREMENT

- *Auscultatory method*: Korotkoff sounds are listened to through a stethoscope placed over the artery being measured during cuff deflation with the pressure reading being shown on a sphygmomanometer.
- *Oscillometric method*: amplitude changes caused by oscillative blood flow against the deflating pressure cuff are measured within an electronic monitor.

The auscultatory method is the system of measuring blood pressure most commonly used by clinicians. The sounds (Korotkoff sounds) generated by the interference of the blood flow against the cuff can be heard in the stethoscope. This method has certain advantages and disadvantages.

One of the most useful advantages is that not only can the blood pressure be determined, but it can also pick up heart arrhythmias (e.g., extra beats). This has been found to be extremely useful in the early diagnostics of cardiac arrhythmia problems. Many hospitals have replaced their electronic oscillometric monitors with the more traditional sphygmomanometers (electronic or mercury column), cuffs and stethoscopes, and use the auscultatory method as the primary method of blood pressure monitoring in wards, outpatients departments and even accident and emergency departments.

The oscillometric method of blood pressure monitoring is found only in electronic blood pressure monitoring devices. It relies not on the sound of the blood pressure but on the transmission of the pressure oscillations signal within the cuff, down the hose to the transducer within the monitor. Because the sensitivity of a transducer is far greater than the human ear, with a stable patient this method offers a greater degree of accuracy.

THE AUSCULTATORY METHOD OF NIBP MEASUREMENT

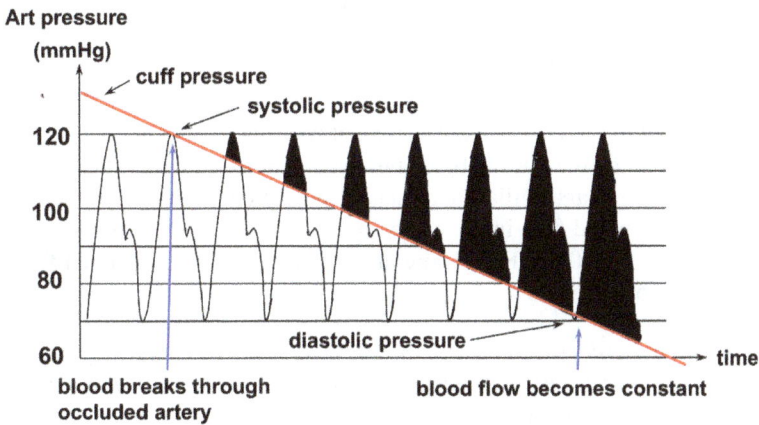

FIGURE 2.3.1 Auscultatory method of NIBP measurement.

What follows is a description of how an auscultatory blood pressure measurement is taken (Figure 2.3.1). The patient is first asked to sit and rest for a few minutes before the measurement is taken. A cuff connected to sphygmomanometer is placed around the patient's left arm. An inflation bulb is then pumped to increase the pressure in the cuff sufficiently to occlude the brachial artery in the arm. For the average adult, this pressure is approximately 160 mmHg. Next the stethoscope is placed on the brachial artery below the cuff. The clinician constantly watches the reading of pressure from the sphygmomanometer as they slowly release the pressure in the cuff by means of a small valve on the inflation bulb.

As the clinician does so, they note when they first heard the sound of blood gushing through the artery against the value shown on the sphygmomanometer. Noting this as the systolic pressure, the clinician then continues to reduce the pressure to such a point that there are no longer any sounds to be heard. At the very instant the sounds disappear; he notes the pressure on the sphygmomanometer as the diastolic pressure.

A more accurate indication of the blood pressure can be obtained by taking three successive blood pressure readings. The first set of readings is ignored to allow for the patient's anxiety over the pressure in the cuff. The next two are summed together and then divided by two to give an average.

The Oscillometric Method of NIBP Measurement

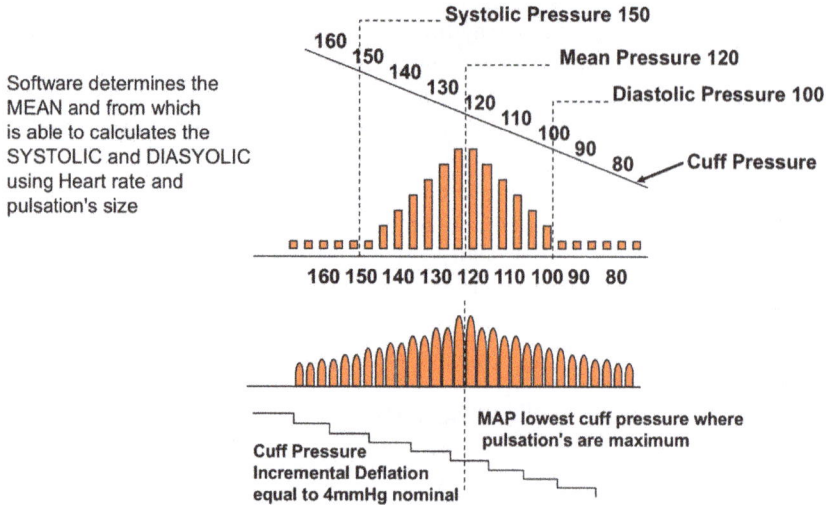

Software determines the
MEAN and from which
is able to calculates the
SYSTOLIC and DIASYOLIC
using Heart rate and
pulsation's size

Systolic Pressure 150
Mean Pressure 120
Diastolic Pressure 100
Cuff Pressure

160 150 140 130 120 110 100 90 80

MAP lowest cuff pressure where
pulsation's are maximum

Cuff Pressure
Incremental Deflation
equal to 4mmHg nominal

FIGURE 2.3.2   Oscillometric method of NIBP.

Oscillometric method of blood pressure measurement relies not on sound but on pressure oscillations generated in the cuff by the arterial blood pumping against the cuff wall. As you can see from Figure 2.3.2, the method appears similar to the auscultatory method. As the pressure is reduced, the interference between the blood flow in the artery and the pressure in the cuff reaches a maximum around the mean pressure (120 mmHg). There are two distinct signals derived from the pressure transducer within the monitor.

1. The first is the cuff's reducing pressure. It may be considered as a steady declining value.
2. The second signal, the transducer picks up, is from the pressure oscillations due to interference of the arterial blood flow and the cuff wall. This may be considered as a varying component.
3. Both signals are fed from the transducer to the microprocessor system that effectively maps both of them within its memory.
4. The microprocessor then determines when the maximum peak-to-peak level of oscillations occurs against the dropping pressure within the cuff. It undertakes to calculate where the systolic and the diastolic occurred.
5. This calculation is based on research that is used within the mathematical calculations, using not only the amplitude of the oscillation and the pressure of the cuff, but also the HR. The mathematical calculation used to determine the systolic and diastolic can be something along the lines of the mean +1/4 for the systolic, and the mean −1/3 for the diastolic.
6. In order for the oscillometric method to work accurately, it is necessary for the patient to have a stable cardiac cycle and not one that includes excessive ectopic beats.

Advantages of NIBP Monitoring

There are many advantages of NIBP.

- It is noninvasive, so there is minimal risk to the patient.
- It is readily obtained.
- It is applicable to adults, children and neonates.
- It is inexpensive.

The points above show why NIBP is possibly the most utilized medical test. It is particularly useful in urgent medical situations where establishing the condition of the patient is imperative. It is also extremely useful in the care of neonates, where it is impossible to determine the blood pressure using the auscultatory method, but an electronic transducer is able to sense the extremely low pressures in the limb of a neonate and thus determine the blood pressure using the oscillometric method.

LIMITATIONS OF NIBP MONITORING

Understandably, there are also some limitations to the use of an NIBP.

- It has limited accuracy in unstable hemodynamic situations (changing blood flow and pressure) such as internal bleeding.
- It is motion sensitive.
- Its accuracy depends on having the right size of cuff available and the correct positioning of the cuff.

A patient whose pulse is varying beat to beat will prove problematic for the oscillometric method of measurement and may even influence the auscultatory method. In this situation, the only reliable measurement of blood pressure would be to use IBP monitoring, which is a direct invasive system only used in operating rooms, intensive care units and high-dependency units.

The right size cuff is a very important factor in taking a correct NIBP measurement. Using electronic monitoring to measure blood pressure you must first ensure that the cuff you select to connect to a limb must be the correct size. Most patient monitors come with a selection of four cuffs. These are often thigh, large adult, standard adult and small adult. If you examine the cuff, you will see that there are two lines marked as the 'Range', and at the end of the cuff is another line marked as 'Index'. When the cuff is wrapped around the limb, you must ensure that the index line falls between the two range lines. You may also notice a small line with a circle on the lower edge of the cuff. This is to indicate that it should be placed over the artery in the limb. Failure to use the right size cuff will often result in readings that are either too high or too low.

A TYPICAL NIBP SYSTEM

Figure 2.3.3 denotes a standard electronic NIBP oscillometric monitor. What follows is the step-by-step operation of such a device when taking a blood pressure.

The Darlington driver is under the direct control of the microprocessor and when the button is pressed to take a blood pressure reading, the Darlington driver is instructed to start the pump and also close all the valves within the system. At this point air is drawn into the pump and fed through the damping chamber 1 to remove the pumps oscillations through Y4 directly up to the cuff. The pressure builds in the cuff, which is then sensed by the transducer.

If an infant cuff is connected, the rising pressure during the first few milliseconds of the pump running will be rapid. The microprocessor is able to senses this from the transducer and thus change the position of Y4, directing the flow to the left through damping chambers 2 and Y3. This second damping chamber is large and thus able to accommodate the additional air that would have gone to an adult cuff. The rise in pressure on the transducer then slows and allows for a more gradual rising pressure in the neonatal cuff. If an adult cuff is connected, the pressure rise on the transducer will be more gradual.

Once the maximum pressure in the system has been achieved, approximately 160 mmHg for an adult or 100 mmHg for a neonate, the microprocessor stops the running of the pump and the air in the system is allowed to settle.

At this point, the microprocessor instructs the bleed valve to open via the Darlington driver. This is driven by a signal that is pulse-width modulated (PWM). In order to drop the pressure in a smooth linear manner, the signal is narrow at first and becomes wider during its operation. At the start of

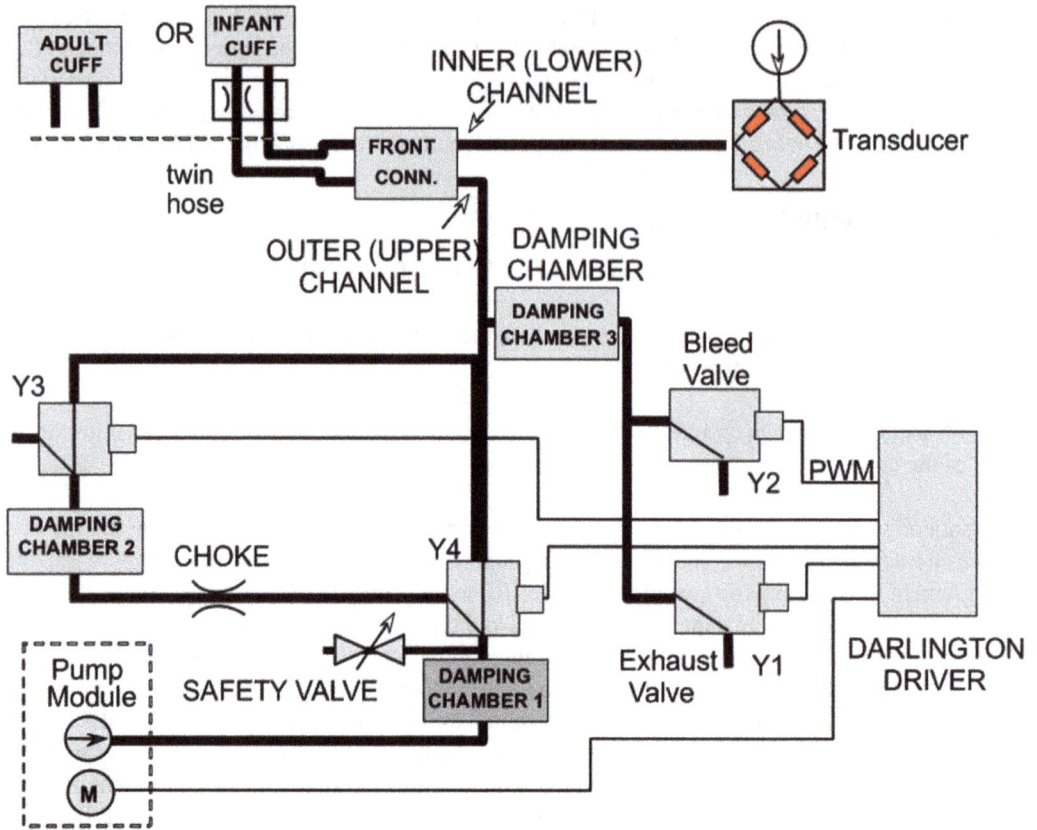

FIGURE 2.3.3 Typical NIBP system.

the deflation, the pressure is high, and therefore opens the bleed valve for a very short period which will evacuate sufficient air. As it proceeds, the pressure in the system is reduced and thus the time the valve is required to be open needs to be longer in order the evacuate sufficient air. The damping chamber 3 is in place to smooth the deflation and remove the oscillations generated by the opening and closing of the bleed valve.

During deflation, the transducer will sense two signals. The first signal is the steady decline in pressure in the cuff. The second signal is of the oscillations generated between the cuff wall and the pulsing artery. These two signals are collected by the microprocessor and mapped into its memory. Once the microprocessor has determined that pulsations seen are not of any significant size of oscillation, it will instruct the exhaust valve Y1 to open. The system is now evacuated of all air and a cuff can be removed from the patient.

At this point, the microprocessor undertakes its calculation of systolic, diastolic and mean and then presents them on the monitor's display.

SECTION 2.4

INVASIVE BLOOD PRESSURE MONITORING

Compared to NIBP, the most obvious advantage of IBP is that it's continuous. Being able to see the readings and the waveforms continuously for a critically ill patient is extremely useful. A critically ill patient's cardiac condition is always vulnerable to a rapid decline at a moment's notice, requiring rapid response and intervention by clinical staff. Unlike NIBP, it does require a greater degree of skill by clinicians in order to use it. It is invasive and therefore comes with some risks.

Monitoring of a patient's blood pressure is critical. Many medical conditions, such as infection and trauma, impact a patient's cardiovascular system. Internal bleeding due to trauma is one such condition and is often seen as the patient's inability to maintain a stable normal blood pressure.

WHY USE INVASIVE BLOOD PRESSURE MONITORING?

IBP monitoring should be used:

- When the hemodynamic status of a patient is unstable.
- During medium and high-risk surgery.
- When repetitive extraction of arterial blood samples is needed.
- When it isn't possible to use noninvasive measurement.

In critical areas and departments of the hospital, such as the operating room (OR) and intensive care units (ICUs), IBP is used routinely. For patients with highly unstable cardiovascular conditions, it is extremely useful for determining the hemodynamic status at an instant. These conditions could be due to trauma or severe disease. During surgery, the patient's cardiovascular condition must be continually monitored along with other parameters by the anesthetist in order to ensure their well-being during the operation.

Invariably, where the cannula is inserted into the patient for the IBP monitoring, there is also attached a three-way port that allows for the extraction of blood for analysis. This eliminates the need for other needles and cannulas to be inserted beneath the patient's skin. Occasionally, there can be a need to use IBP when it is not possible to use an NIBP for reasons such as burns or traumatic injury to limbs.

Patients and their relatives often remark on the bewildering array of number of tubes and cables connected to the patient. This is an understandable reaction when a patient is first seen by relatives in the ICU. Some of this stems from the need to monitor IBP and take regular blood samples.

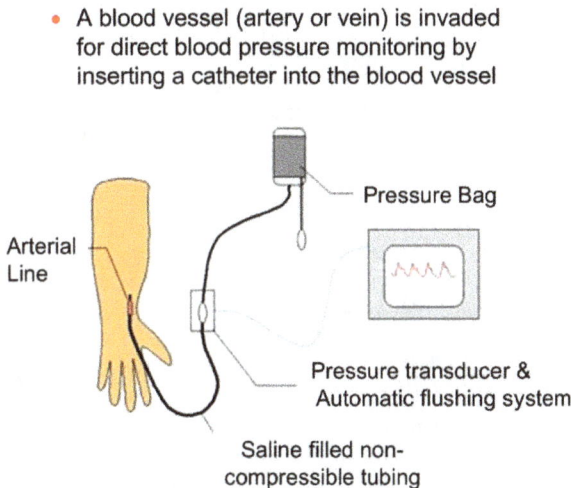

- A blood vessel (artery or vein) is invaded for direct blood pressure monitoring by inserting a catheter into the blood vessel

Arterial Line

Pressure Bag

Pressure transducer & Automatic flushing system

Saline filled non-compressible tubing

FIGURE 2.4.1 Blood pressure measured invasively.

Figure 2.4.1 depicts a typical IBP monitoring set up. The cannula is inserted into the patient's arm with a plastic line, returning to a transducer, an electronic device, which is able to change the pressure felt on its dome into an electrical signal that is fed back to the patient monitor. There will often be a three-way tap connected to it at the cannula. The line then continues up to a plastic bottle/bag which is wrapped by a pressure cuff. Within the plastic bottle/bag and line is a mixture of saline and heparin solution (an anticoagulant that stops blood from clotting).

The cuff surrounding the plastic bottle/bag is pressurized to a high-pressure in the order of 300 mmHg. This ensures that a slow and low-pressure flow of the saline/heparinized solution is forced down across the face of the transducer and through the line into the patient's arm. In the opposite direction, the pressure from the vein or artery that is being monitored will be reflected up through the cannula and along the tubing to the face of the transducer.

If the end of the cannula within the patient's arm gets blocked, then a small roller that is positioned in the line is adjusted to allow for a greater flow of the saline/heparinize solution through the cannula (reduces the viscosity of the blood), and thus removes any blockage at the very tip of the cannula within the patient's arm.

The position of a three-way tap close to the cannula on the patient's arm is adjusted at the start of monitoring to undertake 'zeroing' of the transducer. This tap is turned to block the flow from the patient artery or vein and open the line to room air. As you can see from the Figure 2.4.1, there is a loop that often hangs between the transducer and the patient's arm. This loop contains the saline/heparinized solution and is often not perfectly leveled with a cannula. By opening to air the monitor is able to 'zero out' the effect of the weight of the fluid in the line, this is known as 'zeroing the line'. You will see on the monitor a button for undertaking zeroing of the IBP monitoring line.

COMPONENTS OF AN INVASIVE BLOOD PRESSURE MONITORING SYSTEM

When it comes to the components used to construct an IBP line, it is worth making the effort to see for oneself by arranging to see clinical staff setting up an IBP on the ICU or operating room. When clinical engineers are required to test IBP monitoring, they invariably use an electronic simulator that is plugged into the transducer socket on the front of the monitor.

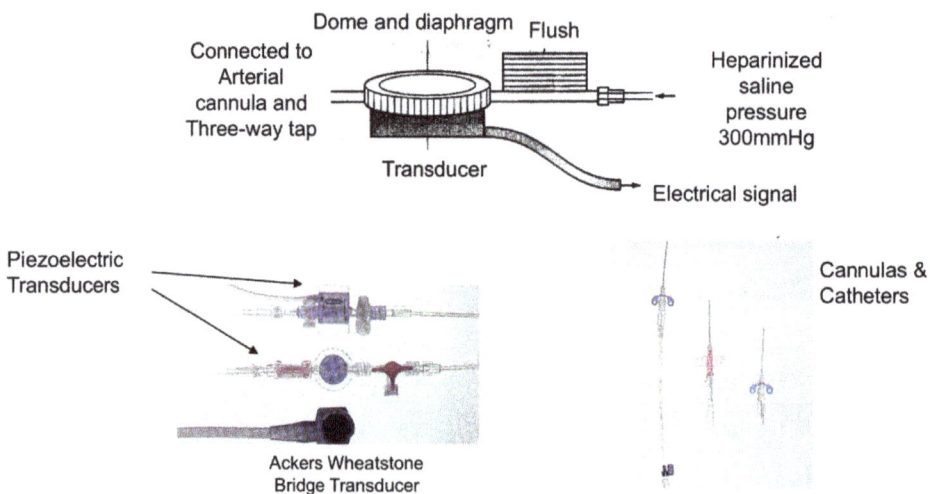

FIGURE 2.4.2   Components of invasive pressure measuring system.

Figure 2.4.2 shows standard connections and components of an arterial line—the three-way tap, enabling blood extraction and zeroing; the transducer, sensing the pressure changes which are fed back to the monitor and displayed and calculated; and the flush adjuster that allows for an extra burst of heparinized/saline in order to remove any blockage from the tip of the catheter.

The figure shows three transducers, two modern piezoelectric transducers which are inexpensive and reliable. The third lower transducer is an older type, Wheatstone bridge transducer, which were used up to around 30 years ago. These were very expensive and very delicate devices. The photograph of the three cannulas shows just some of the differing types available. Each one as you can see has a butterfly that is normally taped to the patient limb to hold it in position.

SETTING UP AN ARTERIAL LINE FOR INVASIVE BLOOD PRESSURE MONITORING

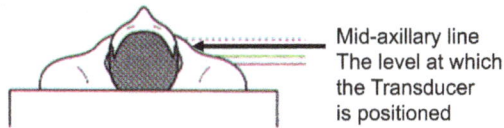

Mid-axillary line
The level at which
the Transducer
is positioned

Zero level has to be established to remove the effects of hydrostatic and atmospheric pressures on the readings

FIGURE 2.4.3   Setting up an arterial line for IBP monitoring.

Figure 2.4.3 shows the importance of aligning the transducer level with the point of monitoring. If the transducer is higher than the cannula, then we will lower the IBP values. Therefore, if the transducer is lower than the cannula, the readings will be higher.

INVASIVE BLOOD PRESSURE MEASUREMENTS

Blood pressure readings throughout the body vary greatly, and what follows is an explanation of various points that can be monitored.

- Arterial blood pressure (ArtBP)
- Central venous pressure (CVP)
- Intercranial pressure (ICP)
- Pulmonary artery pressure (PAP)
- Pulmonary capillary wedge pressure (PCWP)

Listed above are five points for pressure monitoring that can be used. Arterial blood pressure (ArtBP) is often measured on the brachial and radial arteries in an arm and femoral artery in the leg. This is a direct reflection of the output pressures generated by the heart around the systemic system.

CVP is routinely monitored in the ICU as a means of determining the fluid balance within the patient. The catheter is often sited in the neck into the superior vena cava. The pressures are extremely low hovering around or just above or below zero. An increasing pressure can often indicate that the patient is over hydrated. This can be addressed by administering a diuretic to increase urine output. Placement of the catheter for this measurement is often cited in the superior vena cava on the chest close to the clavicle.

ICP within the brain is sometimes measured in order to understand if high or low cerebral spinal fluid pressure is causing certain symptoms. Specialist catheters are used for these types of procedures.

Pulmonary arterial pressures (PAP) and pulmonary capillaries wedge pressure (PCWP) are measured within the heart and the entrance to the lungs, using a very expensive and specialized catheter that is fed from either the inferior vena cava or superior vena cava, through the right side of the heart, to the pulmonary artery (PA), and then into the lungs. It is a procedure that is only undertaken by the most experienced clinicians, often in the operating room. Again, the pressures are extremely low and denote the cardiac output (CO) of the right side of the heart.

LOCATIONS OF INVASIVE BLOOD PRESSURE MEASUREMENTS

Figure 2.4.4 is a hypothetical representation of the circulation system in order to demonstrate the various points that IBP can be monitored.

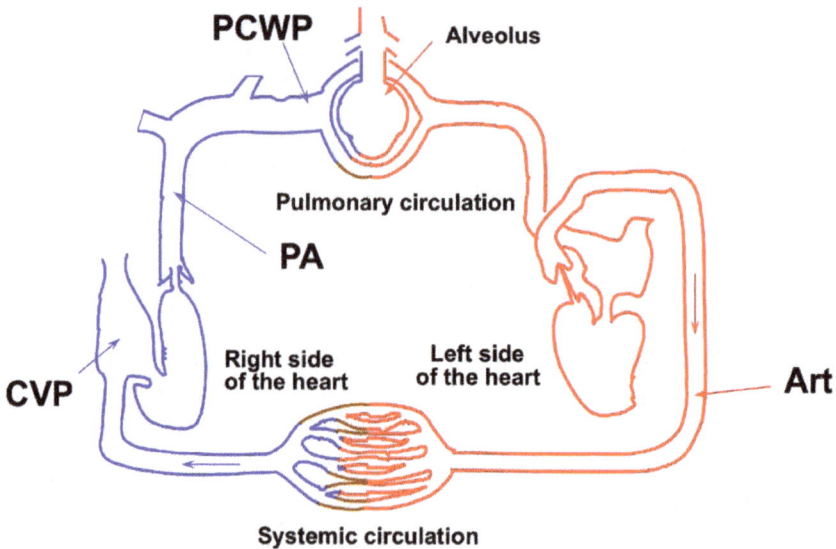

FIGURE 2.4.4   Locations of invasive blood pressure measurements.

Figure 2.4.4 shows the two halves of the circulation system. The first is the pulmonary circulation, the lungs and venous return. The second is a systemic circulation, which is everything else within the body other than the lungs.

Within the systemic circulation there are capillaries that take blood from the arteries as it passes through the various organs and returns by the venous side of the circulation back to the heart. In red are the arterial points of measurement such as the brachial, radial and femoral artery. The blue side of the drawing shows the points of measurement on the venous return to the lungs. This is the CVP, PAP and PCWP. In looking at pressures across the circulatory system, we are able to gain an insight into the cardiac output and systemic vascular resistance.

WHAT DO THE PRESSURES REFLECT?

Figure 2.4.5 explains in more detail the various points that IBP is used to measure. The terms 'afterload' and 'preload' become more significant when clinical staff are undertaking a more detailed assessment of the heart and cardiovascular efficiency using cardiac output (CO) monitoring.

- Art: afterload of *left* ventricle
- PA: afterload of *right* ventricle
- CVP: preload of *right* ventricle
- PCWP: ~preload of *left* ventricle

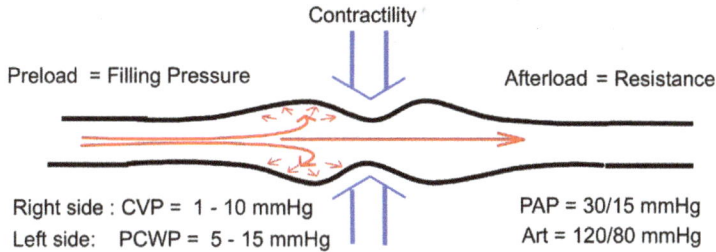

FIGURE 2.4.5 What do pressure reflect?

The terms 'preload' and 'afterload' directly correlate to the position from which the blood pressure is taken. The arterial afterload (ArtBP) is the pressure reflected in the left ventricle, this being the output of the heart to the systemic system. CVP is a preload measured in the vena cava as the blood returns to the right atrium in the heart. The PA afterload is the pressure from the right ventricle as it pumps blood toward the lungs. PCWP is the preload pressure measurement taken in the capillaries of the lungs. As you can see from the values given some of the pressures around and within the heart are extremely low.

The path taken by a catheter that is being used to monitor cardiac output (CO) can be seen in Figure 2.4.6.

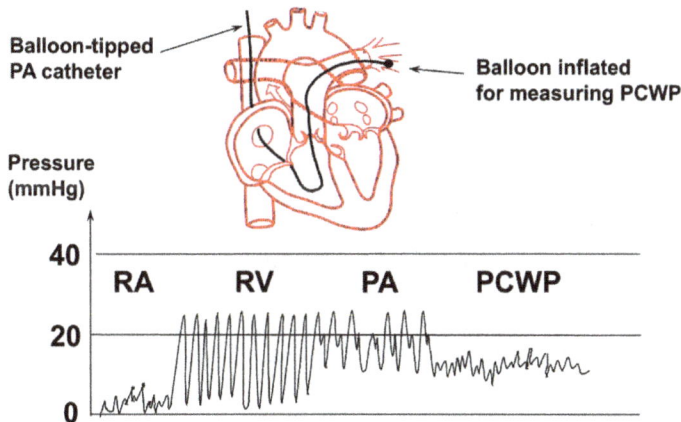

FIGURE 2.4.6 Pulmonary artery catheter.

This shows the path and pressures encountered by a PA catheter that has been inserted via the superior vena cava into the heart. This catheter is sometimes known as a Swan-Ganz catheter. As a clinician guides the catheter through the heart, they are able to ascertain the position of the tip of the catheter by the pressure signal display which is encountered. Once it reaches the capillaries entering the lung, the tip balloon is inflated to block (wedge) the capillary and reflect back the pressure known as PCWP. Arterial IBP is the most common use of IBP and is routinely used in operating rooms and critical care units.

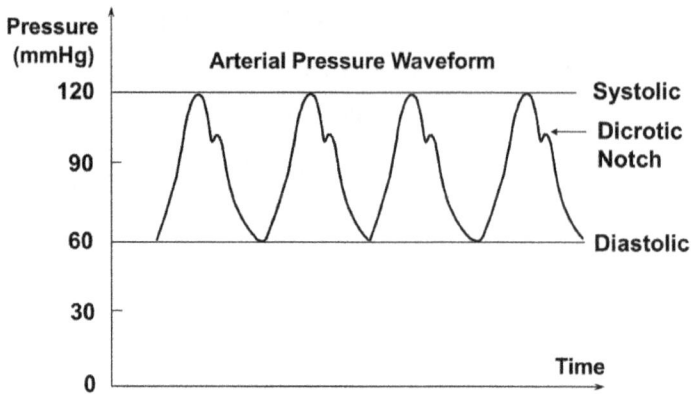

FIGURE 2.4.7 Arterial blood pressure waveform.

Represented in Figure 2.4.7 is what would be expected to be seen on a monitor, measuring the arterial pressure (ArtBP) from perhaps the brachial or radial artery. The dicrotic notch that can be seen is due to the closure of the aortic valve that snaps shut it then sends as it were a small pressure wave on the declining side of the wave form.

Factors Affecting IBP Readings and Displays

Today we see to a great deal less of the problems described in Figure 2.4.8 as most of the IBP lines used come as a complete kit that are designed to reduce 'ringing' in the line.

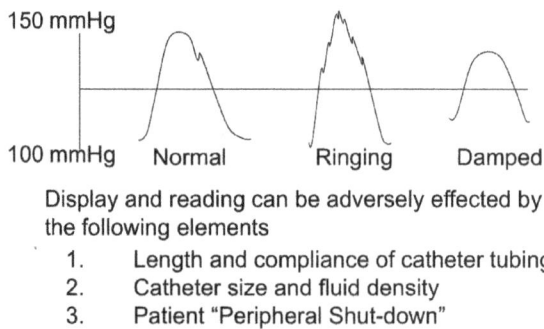

FIGURE 2.4.8 Factors effecting Inv BP reading and displays.

The three factors listed above can each impact on the quality of the blood pressure signal. The catheter tubing length and compliance (elasticity) if not correct will cause 'ringing' in the line as the pressure generates a standing wave within the tubing. This also impacts on the display wave form and readings when a patient becomes cold and goes into peripheral shutdown reducing the blood flow.

ADVANTAGES AND DISADVANTAGES OF IBP MONITORING

To summarize, the chief advantages of IBP are as follows:

- Blood pressure is measured directly.
- Blood pressure is measured continuously.
- IBP provides fast and accurate information.
- It provides for the convenient extraction of blood samples.

Set against these advantages are some disadvantages:

- IBP is expensive, especially the PA catheter.
- There is a risk of infection.
- There is a risk of other complications.
- It requires well-trained clinical staff.

Anything invasive always carries a degree of hazard and IVBP is no different.

POSSIBLE CAUSES OF UNRELIABLE IBP READINGS

IBP reasons can be rendered unreliable by a number of possible causes:

- incorrect zero level
- air in the fluid line
- partial or total occlusion of the catheter
- loose system connections
- a broken transducer

As stated before, the use of IBP requires highly skilled clinicians and a great deal of care. The reasons for unreliable readings stated above cannot be overstated. Air in the line is a potential hazard to the patient if it migrates and enters the blood stream.

SECTION 2.5

TEMPERATURE MONITORING

Knowing a patient's temperature is one of the most important parameters we have for understanding a patient's general well-being. In infancy, mothers will often check a child's temperature by simply placing their hand on the child's forehead. With a great degree of accuracy, the mother is able to determine whether a child has a fever. A rising temperature in any patient inhibits the ability of bacteria and viruses to cause infection and is of course part of the patient's defense mechanism. Countless medical conditions are associated with a rise in the body temperature, including COVID and the common cold. The organ within the body most associated with producing heat is the liver, the largest organ in the body. This organ can possibly be thought of as a chemical factory, and, of course, in almost all chemical reactions there is the production of heat. All the other organs, tissues and muscles also produce heat. A term most commonly associated with temperature is 'metabolism'. This is the consumption of oxygen along with proteins, fats and carbohydrates that produce growth, energy and fight infection. This heat is transferred by the cardiovascular blood flow and circulated around the body. So, blood flow plays a vital part in transferring heat.

TEMPERATURE MONITORING DURING SURGERY

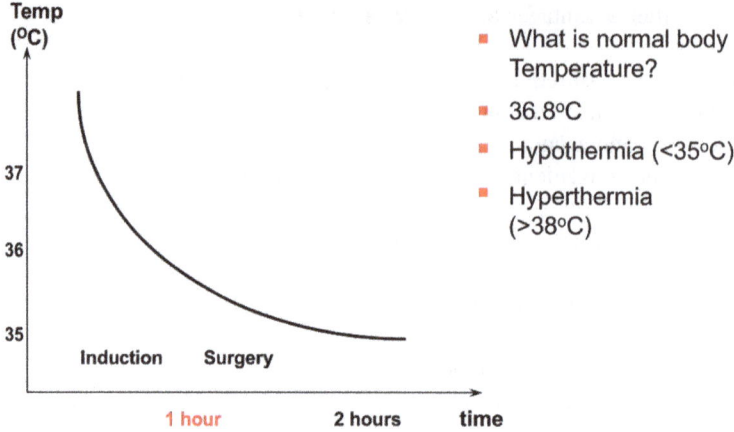

FIGURE 2.5.1   Temperature monitoring during surgery.

Figure 2.5.1 shows the typical temperature drop for a patient undergoing either thoracic (Chest) or abdominal (Stomach) surgery. As can be seen in Figure 2.5.1 with a normal body temperature of 36.8°C, a patient is usually brought to the operating room in a very light theater gown, they will begin to lose heat and thus their core temperature will begin to steadily drop. Following induction, when the patient is put to sleep and surgery itself begins, again the patient's core temperature will fall. Once the patient has been surgically opened and the internal organs no longer have the insulation properties from the skin further heat loss is increased.

The core temperature of an individual is usually understood to be 36.8°C. This is known as normothermia. If the patient's temperature were to drop below 35°C, we have a condition called hypothermia. Again, if the patient's condition caused a rise in temperature above 38°C, we then have the condition hyperthermia (Figure 2.5.2). Please remember that this is the core temperature within the chest or abdomen of the patient and not a peripheral temperature or skin temperature. It should also be pointed out that temporal (brain) temperatures are often slightly higher than core temperatures as the brain has a very high metabolic rate.

There are different ways of monitoring the temperature and it depends on the type of operation and the type of anesthesia used as well as on the medical condition and age of the patient. Intraoperative (induction and surgery) temperature monitoring is used in instances such as long operations, extensive abdominal, thoracic and neurosurgery. During intraoperative monitoring, a fall in temperature is to be expected. During postoperative (recovery) monitoring, temperature should show a gradual rise in the patient leading back to normothermia (a normal temperature).

HYPOTHERMIA AND HYPERTHERMIA

FIGURE 2.5.2   Hypo- vs. hyperthermia.

The conditions of hypothermia and hyperthermia are always indicators of a change in a patient's medical condition and must be acted upon in order to bring it back to normothermia. Hyperthermia is often caused by infections and diseases and requires constant monitoring because prolonged periods of this condition can cause long-term health problems.

Malignant hypothermia is a rare condition that can happen when a patient's reaction to the anesthetic vapor causes a hypermetabolic response. This is potentially a life-threatening condition and requires immediate action by the anesthetist. The first indication of this condition is a sudden rise in the carbon dioxide (CO_2) expiration value from the patient. This is then confirmed by a rapid rise in the patient's temperature.

Hypothermia is often encountered when a patient has just been admitted to hospital and may be suffering from exposure. Here again, the aim is to return the patient back to normothermia. In the event of the patient suffering trauma, particularly neurological trauma, great care is taken to raise the patient's temperature in a gradual manner rather than a sudden warming. There are occasions particularly within the operating room when the patient's temperature is reduced deliberately. This offers the surgeon the opportunity to work on tissue with a reduced blood flow, and thus enable increased clarity when undertaking suturing (stitching).

There are two different types of hypothermia that are of interest:

- *Intentional*: during neurosurgical or cardiac procedures, the blood flow and oxygen consumption are deliberately decreased.
- *Inadvertent*: the patient loses heat because of factors such as the temperature in the operating room, for instance from air-conditioning, inspiratory gasses from wall outlets and cylinders, and receiving cool liquid infusions.

There are also two different types of hyperthermia that are of interest:

- Predisposed hyperthermia (malignant hyperthermia), which is triggered by anesthetic agents. Symptoms can include muscular rigidity, hypertension, skin mottling. When temperature rises above 42°C, the mortality rate (death) is 80%–90%.
- The second reason for hyperthermia can be a fever with no time to treat an infection before operating in emergency cases. Neurological diseases and conditions may also alter temperature regulation in the body. This is controlled by an area of the brain call the hypothalamus.

INDICATIONS FOR PERIOPERATIVE TEMPERATURE MONITORING

Interoperative period

- Long operations
- Extensive abdominal or thoracic surgery
- Surgery with induced hypothermia
- Infants and young children

Postoperative period

- Shivering and cold patients

A patient who has become profoundly cold during surgery will need to be closely monitored during recovery. As the patient became cold, the anesthetic agent in their blood circulation may have become partially trapped in their peripheries, such as the legs and arms. As the patient is warmed in the recovery area, this anesthetic agent can re-enter the blood flow due to vasodilation (blood vessels begin to open wider) and can in rare situations cause a delay in the recovery from the anesthetic agent.

CONSEQUENCES OF PERIOPERATIVE HYPOTHERMIA

Consequences can include the following:

- Decreased oxygen uptake and CO_2 production (decreased metabolic rate)
- Shivering—increased oxygen consumption
- Changes in HR and rhythm
- Decreased cerebral blood flow

The points above demonstrate the importance of avoiding perioperative hypothermia. The reduction in the metabolic rate is clearly unwelcome and can impact the heart and its rhythm. Decreased blood flow to the brain also impacts many of the autonomic controls the brain has over other organs. Postoperative shivering due to prolonged hypothermia is a negative consequence causing increased oxygen demand.

CONSEQUENCES OF PERIOPERATIVE HYPERTHERMIA

Consequences can include the following:

- Increased oxygen uptake and CO_2 production (increased metabolic rate)
- Fever
- Excessive surgical draping
- Malignant hyperthermia
- Septicemia

Perioperative hyperthermia is, of course, a situation that should be avoided. Often considered to be a temperature of 37.5°C and above it can be due to the failure of thermal regulation by the hypothalamus within the brain. Septicemia (blood poisoning) is a condition caused by bacterial infection and their toxins and always leads to a rapid rise in temperature. Consequences of perioperative hyperthermia are increased oxygen demand, respiratory work and cardiac work.

MONITORING TEMPERATURE

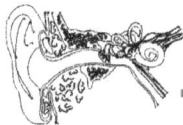

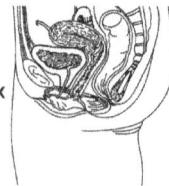

FIGURE 2.5.3   How to monitor temperature?

Figure 2.5.3 lists a few of the most common sites on the human body for which we are able to monitor the patient's temperature. Tympanic is routinely used by clinicians today by means of a tympanic infrared digital monitor. There are however some drawbacks to this method of temperature measurement. Firstly, the plastic cover covering the tip of the sensor must be changed with each use to ensure that the infrared signal is not disrupted by earwax. Secondly, the infrared signal must be accurately positioned within the ear in order to ensure maximum signal reflection from the inner membrane of the ear.

A more recent development has been the temporal artery thermometer. These devices use a similar infrared technology to the tympanic infrared thermometer. The method of measurement used is to scan across the forehead along part of the external carotid artery above the left eye. The external carotid artery at this point on the forehead is just 1 mm below the skin surface and thus a relatively accurate reflection of the arterial blood temperature.

For esophageal (mouth), nasopharynx (nose), rectum (anus) and axillary (armpit), a dual thermistor bead sensor at the end of a cable is used to monitor these sites. Thermistors (heat sensitive resistors) are encapsulated within the bead tip. This approach to temperature monitoring is generally used for continuous long-term temperature monitoring. Esophagus or nasopharynx reflects well the overall changes in the body core temperature. Skin temperature reflects the peripheral temperature being influenced by ambient environmental temperature and/or peripheral blood flow.

TEMPERATURE MONITORS

POSTOPERATIVE TEMPERATURE MONITORING

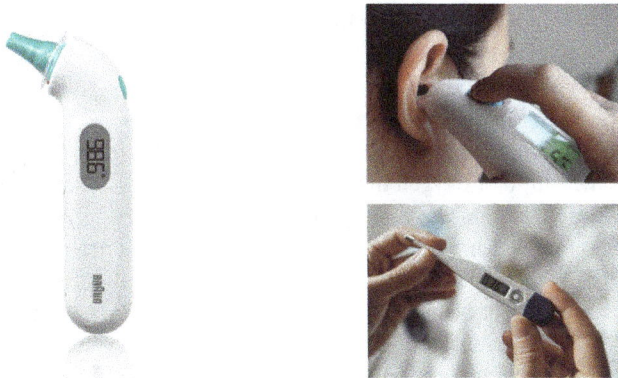

FIGURE 2.5.4   Temperature monitors.

It can be beneficial to monitor postoperative temperature at two sites (Figure 2.5.4).

- Parts of the body may markedly differ in temperature from each other.
- Monitoring of temperature differences (dT) at two sites (for instance, the esophagus and peripheral) provides indirect information on blood flow.

As discussed previously, temperatures across the human body can vary substantially. The temperature on the skin on an extremity such as a hand or a foot may be as much as 2°C different from the core temperature. Measuring temperature gradient across to sites, such as a skin graft, can give a reasonable indication of blood flow and prove useful in determining the success of the graft. The rate at which temperature difference of the two sites, changes provides indirect information on general blood flow as well as is helpful in observing against overshooting of warming and cooling a patient.

SECTION 2.6

SATURATED PULSE OXIMETRY (SpO$_2$) MONITORING

The advent of saturated pulse oximetry is probably the single most important advancement in patient monitoring in the last 40 years. I remember well the very first time I was handed an SpO$_2$ monitor: I was attending a service engineering meeting for the company I worked for and was handed a small monitor with a finger probe. Not only did it give the HR, but also a new number representing the oxygen level in the arterial blood flow in the particular finger the sensor was fitted to. My then-manager asked each of us what we believed the sales price of such a device might be. Given that at that time a large defibrillator might sell from around 5000USD, we all estimated the price to be in the range of 1000USD. The manager smiled and informed us that the company was about to sell this new monitoring parameter and the sales price would be around 5000 USD. For such a small device, the price seemed extortionately high, and its significance was yet to be appreciated. I now look back when I see SpO$_2$ monitors available for sale in such places as pharmacies for around 50USD and remark on just how much we have come to rely on SpO$_2$ monitoring.

As many textbooks will tell you, the father of pulse oximetry is a Japanese biomedical engineer, whose name was Takuo Aoyagi, who worked for Nikon Kohden in Japan. Earlier work had been undertaken in Germany by physician Karl Matthes and also by British and American military physicians. Low blood oxygen was recognized as a problem for pilots of military aircraft at high altitudes. One description of the research that was undertaken was that by adding to the pilot's helmet, a red transmitter and detector, across one ear, and an infrared transmitter and detector across the other ear with readings being sent to a dual channel chart recorder beneath the pilot's seat. A cannula was also inserted into the pilot's arm with a line down to a collection jar also beneath his seat for the collection of a sample of blood at high altitude. Once the required altitude was reached by the plane, the pilot would open the cannula and take a blood sample. At the same time, a dual-channel chart recorder would record the levels of red and infrared light received by the detectors in the helmet. Upon returning to the ground, chart recordings and blood samples would be taken to the laboratory and a correlation calculated between the received signals on the chart recorder and the arterial blood gas measured of oxygen in the blood sample. This, of course, was an arduous and laborious process that might be considered impractical.

Takuo Aoyagi's invention was in the development of a single-digit probe (finger probe/ear probe) that contained both the red and infrared LEDs and the red/infrared detector. Central also to this technology was the recent advent of microprocessors and LEDs, which enabled real-time calculations of both red and infrared signals and the calculation of the received signal level ratios. Within a very short period of time, the modern pulse oximeter was developed further and became widely available during the early 1980s. Almost immediately, it was accepted in the operating room as a minimum monitoring standard and thus ensured the patient's oxygen was continually monitored during anesthesia.

For a conventional pulse oximeter, one needs to understand that the reading displayed is a relative and not an absolute value. In practice, only a blood gas analyzer will give you the absolutely accurate reading of oxygen in the blood. This can be considered as the gold standard reading. Though a pulse oximeter will closely follow the value of a blood gas analyzer, but due to reasons that will be explained later, its accuracy is not perfect, and it is therefore to be thought as a relative value. Its major advantage over an arterial blood gas reading of oxygen is that it is continuous and easy to use and there is, of course, no invasive need to sample blood.

UNDERSTANDING OXYHEMOGLOBIN

Oxyhemoglobin—Oxygen that is bound to hemoglobin (red blood cells).

- Each haemoglobin molecule consists of 10,000 atoms, four of which are iron atoms that attract and hold Oxygen.
- Each red blood cell contains about 250 million haemoglobin molecules
- Each individual has approx. 5,000 cc's of blood, and each cc contains 5 billion red blood cells

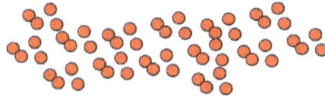

FIGURE 2.6.1   Understanding oxyhemoglobin.

Blood is made up of four elements: plasma, white blood cells, platelets and red blood cells (hemoglobin). Hemoglobin itself can be divided into two: deoxyhemoglobin (hemoglobin low in oxygen and is dark red in color) and oxyhemoglobin (rich in oxygen with a bright red color). The details in Figure 2.6.1 demonstrate just how little oxygen each hemoglobin molecule carries, and the large volume of hemoglobin required to carry sufficient oxygen for the body's needs. Not only does the hemoglobin molecules carry oxygen, but they can also carry the CO_2 (produced by metabolism) and other gasses such as carboxyhemoglobin.

ESTIMATING TISSUE OXYGENATION

Methods for Estimating Tissue Oxygenation

- ABG - Arterial Blood-gas analysis: determines PO_2 (partial pressure)
  - PaO_2 = Partial atmospheric pressure of oxygen dissolved in Arterial Blood
  - If 1 Atmosphere is 760 mmHg, and contains 21% oxygen which accounts for 160 mmHg in room air
  - A usual reading in a healthy individual is around 90-106 mmHg

FIGURE 2.6.2   Methods for estimating tissue oxygenation.

Arterial blood gas analysis is often referred to as the 'gold standard' and is considered the most accurate method of measurement of oxygen in the bloodstream (Figure 2.6.2). Using an instrument known as a blood gas analyzer, a chemical method of estimation is used to give the most accurate level of partial pressure oxygen (SaO_2). To understand partial pressure, you have to first consider that air consists of approximately 79% nitrogen and 21% oxygen. There are, of course, small traces of other gases such as CO_2 and argon. At sea level, 1 atmosphere will give a barometric pressure of 760 mmHg, and 21% of that pressure is oxygen, so oxygen is therefore said to have a partial pressure of 160 mmHg (about 1/5th of the total barometric pressure). Within the arterial blood flow, the SaO_2 pressure in the human body is around 90–106 mmHg and can be said to be the partial pressure of oxygen concentration in the arterial blood flow.

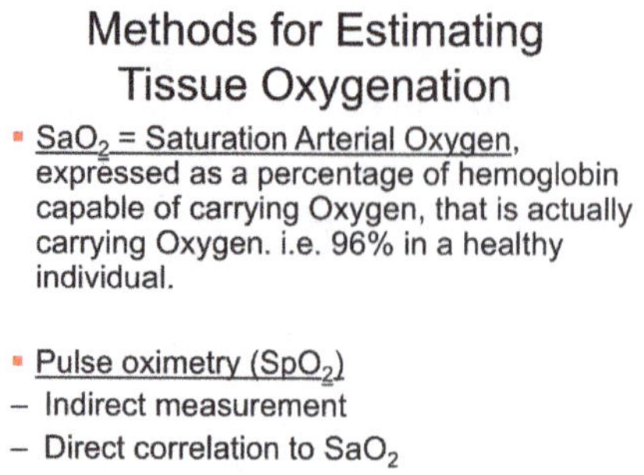

Methods for Estimating Tissue Oxygenation

- SaO_2 = Saturation Arterial Oxygen, expressed as a percentage of hemoglobin capable of carrying Oxygen, that is actually carrying Oxygen. i.e. 96% in a healthy individual.

- Pulse oximetry (SpO_2)
  - Indirect measurement
  - Direct correlation to SaO_2

FIGURE 2.6.3   Methods for estimating tissue oxygenation.

Not all hemoglobin will at any point be carrying a full load of oxygen. Some of the hemoglobin may be dysfunctional (not capable of carrying gas) and some may also be carrying other gasses, particularly CO_2. The term SaO_2 (saturated arterial oxygen) is the value shown on a blood gas analyzer that has measured the value in the blood sample. Comparing SpO_2 to SaO_2 (Figure 2.6.3), there is a direct but an important difference 'P' in the term SpO_2, which denotes that it was derived from a pulse oximeter that relies on an arterial blood flow showing a pulsatile arterial waveform (a constantly varying blood pressure wave).

HEMOGLOBIN

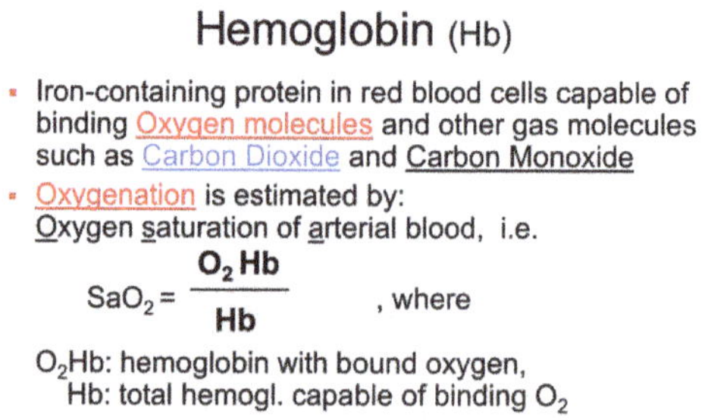

Hemoglobin (Hb)

- Iron-containing protein in red blood cells capable of binding Oxygen molecules and other gas molecules such as Carbon Dioxide and Carbon Monoxide
- Oxygenation is estimated by:
Oxygen saturation of arterial blood, i.e.

$$SaO_2 = \frac{O_2\,Hb}{Hb}, \text{ where}$$

O_2Hb: hemoglobin with bound oxygen,
Hb: total hemogl. capable of binding O_2

FIGURE 2.6.4   Hemoglobin (Hb).

Within the hemoglobin, red cells there are iron-containing proteins capable of binding with oxygen, CO_2, and carbon monoxide molecules. The ability of each gas to attach to these iron-containing proteins is variable. This is called gas affinity. Carbon monoxide has the strongest affinity to attach to the iron-containing proteins followed by CO_2 and lastly oxygen. The small formula in Figure 2.6.4 shows how the calculation of saturation is achieved. It is oxyhemoglobin divided by total hemoglobin which results in the SaO_2 percentage.

HEMOGLOBIN (Hb) AND SpO_2

- Oxyhemoglobin (HbO_2) is bright red
- Deoxyhemoglobin is dark red

Pulse oximetry makes use of the fact that Hb (deoxyhemoglobin) and HbO_2 (oxyhemoglobin) have different light absorption properties. The color of arterial (oxygenated) blood is bright red while the venous blood is darker. Two light beams, red at 660 nanometer and infrared at 940 nanometer wavelength, are shone through the finger/ear and partially absorbed by tissue, bone, venous blood and the nonpulsatile arterial blood. A varying level of light absorption is caused by the pulsatile flow caused by the regular beating of the heart. This dynamic part of the signal is called the plethysmogram and is used to determine the SpO_2 saturation. The amplitude of the plethysmograph is a measure of the perfusion (volume of blood flow). It can be displayed as a wave form or by means of a signal strength indicator.

LEVELS OF SaO$_2$ IN CIRCULATION

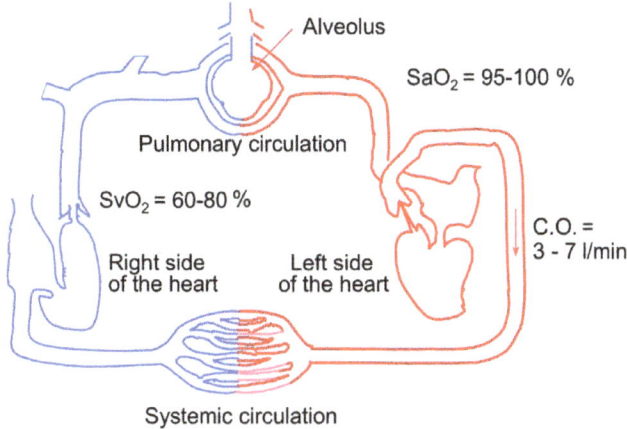

FIGURE 2.6.5   The levels of SaO$_2$ in circulation.

Figure 2.6.5 represents the two sides of the circulatory system of the human body and has been divided into red and blue. Red is the arterial blood flow around the body that then passes through the systemic circulation. The systemic circulation is made up of the other major organs of the human body such as the liver, kidneys and the brain. Within these organs, oxygen is consumed from the arterial blood flow and CO$_2$ produced.

Flowing out of the systemic circulation is the venous return to the heart and lungs. This blood is lower in oxygen but rich in CO$_2$. As you can see arterial blood is denoted as SaO$_2$ and venous blood as SvO$_2$. CO referrers to cardiac output flow in liters per minute for an adult. Returning blood from the systemic circulation enters the right side of the heart and then back to the pulmonary circulation (lungs). CO$_2$ is breathed out and the intake of air which contains a high level of oxygen is allowed to be absorbed back into the arterial blood flow.

CONSTRUCTION OF AN SpO$_2$ PROBE

- Red Light = ~ 660 nm
- Infrared = ~ 910 nm

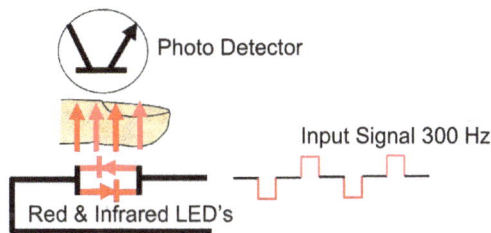

FIGURE 2.6.6   SpO$_2$ light source and detector.

The construction of an SpO$_2$ probe is a fairly simple affair (Figure 2.6.6), containing two LEDs and a photo transistor detector. Each of the LEDs transmits light of a very accurate wavelength. For red, this is around 660 nm wavelength, and for infrared, 910 nm wavelength These two LEDs are wired back-to-back and then a signal of approximately 300 Hz is applied across the LEDs, which will then alternate between on and off.

When the positive phase of the signal is applied across the infrared LED, it will light. When the opposite negative phase of the LED is applied, then the red LED will transmit light. Between each phase is a dark phase when neither LED is on. This is important as it allows the photodetector/microprocessor to account for any ambient light that may have fallen onto the detector from an outside source such as daylight. This ambient light signal is then deducted from the signals of red and infrared in order to obtain just the directly received light of each.

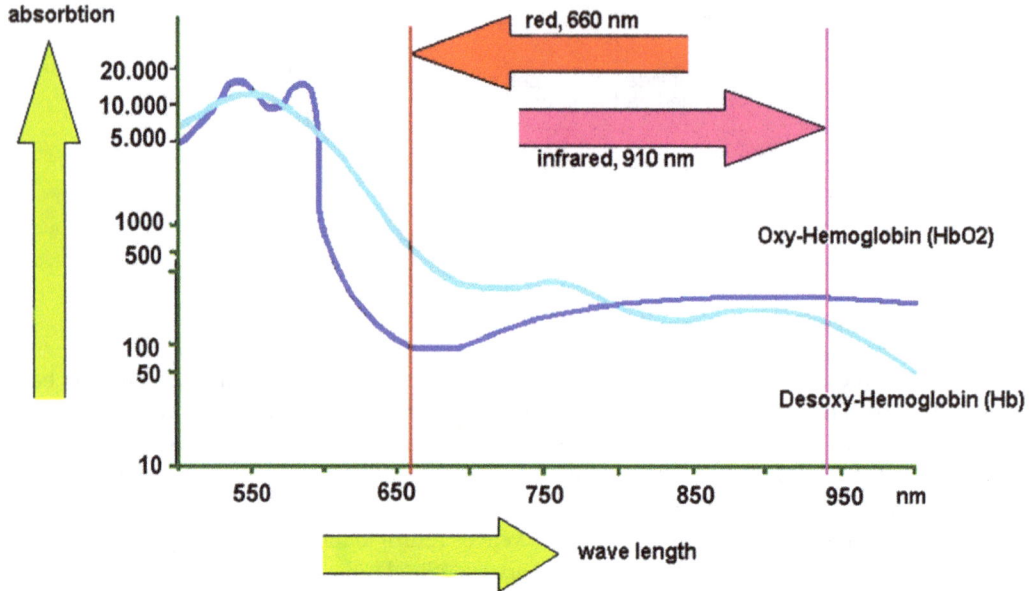

FIGURE 2.6.7 Intra red- and red-light absorption within hemoglobin.

Figure 2.6.7 demonstrates two wavelengths of light used in SpO_2—the red line being the visible red LED light at approximately 660 nm and the purple line of infrared LED light at approximately 940 nm. Oxyhemoglobin is represented by the blue curve line and deoxyhemoglobin is represented by the cyan curved line.

The level of absorption of the red light through the oxyhemoglobin (bright red) is considerably less than the absorption of red light through the deoxyhemoglobin (dark red). When looking at the levels of absorption at 940 nm infrared light, it may be surprising, but the absorption is reversed. Here, infrared light finds it more difficult to pass through oxyhemoglobin than deoxyhemoglobin. It is this characteristic of the two levels of absorption, red and infrared with regard to oxyhemoglobin and deoxyhemoglobin, that enable us to calculate the level of oxygen in the arterial blood flow.

For each level of saturation between 80% and 100%, there is a ratio of red/infrared absorption that is used along with information on the volume of the dynamic arterial pulse to calculate the level of SpO_2.

MEASUREMENT

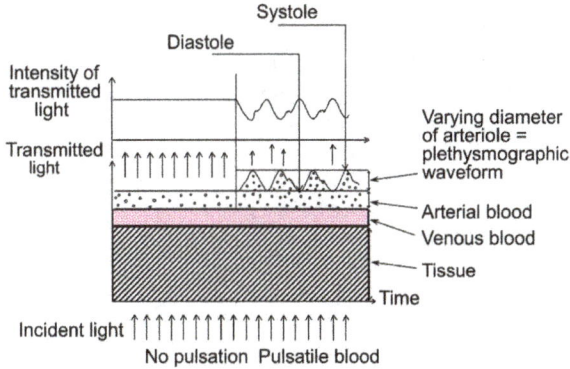

FIGURE 2.6.8   Measurement principle.

If you examine Figure 2.6.8, you will see that the vast majority of the light transmitted is absorbed by tissue, venous blood and the static arterial blood. The pulsatile arterial blood, which is the dynamic element needed, is plotted against time in order to indicate the volume of blood beneath the curve.

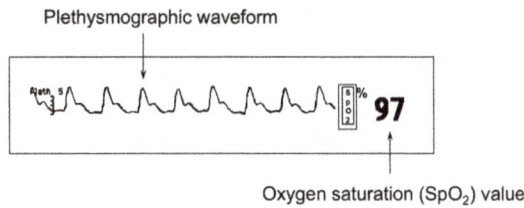

• **Integrity of waveform ensures reliability of SpO$_2$ value**

FIGURE 2.6.9   SpO$_2$ monitoring.

An **SpO$_2$** monitor with a display is particularly useful in not only calculating the level of oxygen in the arterial blood (Figure 2.6.9), but also in understanding the patient's peripheral perfusion, which is the volume of blood passing through an artery. This is very important when fully understanding peripheral circulation and the reasons for poor perfusion. The pulse/plethysmographic waveform amplitude will be continually monitored by clinicians to ensure adequate peripheral perfusion of the patient. If the patient becomes cold or goes into peripheral shutdown (vasoconstriction), then intervention may be required to restore adequate perfusion. The plethysmographic waveform can be thought of as a 'parameter' in its own right such is its importance.

SpO$_2$ AND NEONATES

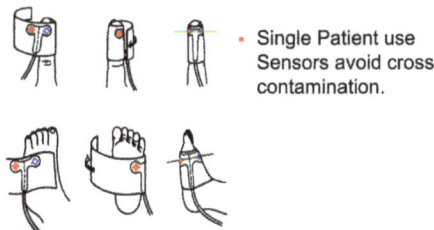

• Single Patient use
Sensors avoid cross
contamination.

FIGURE 2.6.10   Neonatal application of SpO$_2$ probes.

Normally applying a finger probe or even an ear probe to a child or adult is a very simple and uncomplicated procedure. But when it comes to applying a probe to a neonate, either on the hand, foot or ear, it is somewhat more difficult and requires a clinician to take time to correctly align the LED transmitters with the photodetector (Figure 2.6.10). There is also a very great risk of applying too much tension across the probe and thus constricting the very signal they are trying to measure. Of course, it should not be too loose in order to avoid misaligning sensor and LEDs due to the movements of the neonate.

The Benefits of SpO$_2$ Monitoring

SpO$_2$ monitoring is as follows:

- Fast
- Continuous
- Inexpensive
- Uncomplicated
- Noninvasive
- Suitable for transport

The factors above explain the widespread adoption of SpO$_2$ monitoring. In particular, the ability to continuously monitor a patient's oxygen can be crucial during a medical crisis. The cost of the monitor and its probes will quickly prove advantageous against the cost of routinely taking arterial blood gases to measure oxygen. Being noninvasive is also a great advantage. When being used for transport, it is a useful parameter but can be unreliable due to transport vibrations. The vibrations can cause a movement in the static arterial and venous blood flows which will inevitably lead the SpO$_2$ monitor to miss calculate and read falsely low.

Limitations of SpO$_2$ Monitoring

Despite its advantages, SpO$_2$ has limitations in some scenarios and can be a 'fair weather friend'.

- Dysfunctional hemoglobin
  - Anemia (low blood count)
  - Carboxyhemoglobin (carbon monoxide poisoning)
  - Heavy smoking can exacerbate difficulties
  - Medical dyes can interfere
- Low perfusion
  - Shock, hypothermia and medications
  - Skin pigment and nail varnish can interfere
    - Manufacturers now ensure that their devices work accurately over all skin tones from white to black

The term 'fair weather friend' alludes to the fact that a conventional SpO$_2$ meter is extremely useful in giving a reasonably accurate reading when the patient saturation is good. However, it is not so good when the patient saturation falls very low, usually below 90%. In this condition, the accuracy against an arterial blood sample will vary by as much as 4% or 5%. With the exception of one particular model, a pulse oximeter produced by Massimo, which uses a revolutionary method of calculation called SET technology (signal extraction technology), most conventional pulse oximeters have their limits.

As was discussed in the first slide, SpO$_2$ is a relative value, not an absolute, so it is to be expected there can be slight differences between a blood gas analyzer (ABG) and an SpO$_2$ reading of saturation. Listed above are a number of reasons why SpO$_2$ is not perfectly accurate. Anemia, a low blood cell count is one of the most common reasons for this. Smoke inhalation can cause carboxyhemo-globin poisoning and again will substantially affect the SpO_2 reading. A blood sample taken from a patient with carbon monoxide poisoning will be seen to be extremely light red in color and yet this is not due to a high level of oxygen, but rather the effect of carbon monoxide molecules. Within the hospital settings, patients are often administered medical dyes prior to such procedures as CT scans and again this will affect the color of the blood and absorption of both red and infrared light.

Patients experiencing low profusion will also make it extremely difficult for a conventional pulse oximeter to accurately accumulate sufficient signal levels for SpO_2 readings, and thus may not be able to give an SpO_2 value. In the early years of SpO_2, there was some difficulty in obtaining accurate readings from people with black or very dark skin, but with the development of much improved LED technology this is no longer such an issue. However, it is often overlooked by medical staff when applying finger probes that the patient may have applied nail varnish to the fingers, and of course this will greatly reduce the signal from the detector as a great deal of light is absorbed by the varnish.

Other complications can include the following:

- Motion artifacts
  - Transport
  - Shivering
- Inadequate blood flow
  - From blood pressure cuffs
  - And tight clothing
- Electrosurgery
  - Diathermy
- External light
  - For example, theater lights

Other reasons for being unable to obtain satisfactory readings are listed above and can be easily avoided, such as not placing a blood pressure cuff on the arm that the SpO_2 probe is connected to. Tight clothing, including wrist watches should be removed, the presence of extremely bright operating room lights in the operating room and diathermy can also severely impact on the SpO_2 monitors ability to give stable readings.

SpO_2 Disassociation

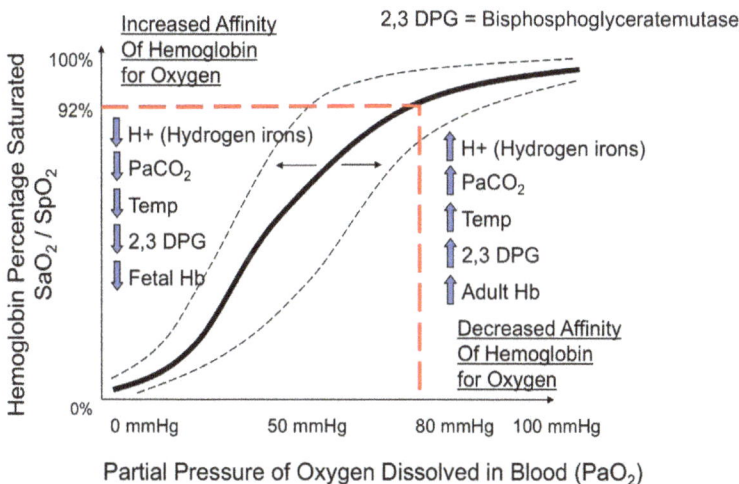

FIGURE 2.6.11 SpO_2 disassociation curve.

Figure 2.6.11 demonstrates the effect of various parameters that can affect the accuracy of SpO_2. Hydrogen ions (H^+) directly reflect the acidity or alkalinity of the blood and are associated with the level of CO_2 in the bloodstream. $PaCO_2$ (partial pressure of CO_2) directly impacts the ability of hemoglobin in respect to carrying oxygen as CO_2 molecules will be attached to the hemoglobin cells in place of oxygen molecules. The effect of a rise in temperature is that it causes an expansion of the blood fluid and thus fewer hemoglobin molecules beneath the pulsatile curve as the arterial blood temperature increases. 2,3-DPG impacts on the ability of hemoglobin to acquire and hold oxygen and CO_2 molecules also impacts the gas affinity of hemoglobin.

During the first few days and weeks of life of a neonate, hemoglobin has substantially more ability to carry oxygen, and this can produce false readings of SpO_2 when monitors are set to adult mode not neonate mode. Many modern monitors are capable of being changed to neonatal mode or may be able to select neonatal mode when the monitor recognizes that the neonatal sensor is attached.

It is worth reiterating that SpO_2 is a relative value and not an absolute figure.

3 Defibrillation

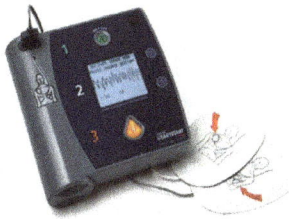

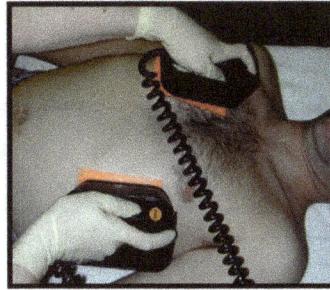

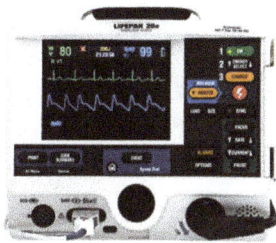

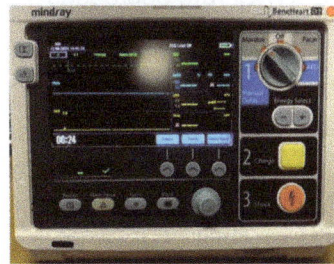

Claude S. Beck
and his first
defibrillator
Courtesy of
Allen Memorial
Medical Library,
CWRU

Note: In order to fully understand defibrillation, it is required that you have studied the sections within this book on 'Circulation and the Heart' and 'ECG'.

Defibrillation is one form of therapy that appears to run counter to everything we understand about the effects of electricity on the human body. In normal circumstances an electrical shock, with high voltages and high currents will have a severe negative impact, causing tissue burns, organ damage and often death. However, in a critical event such as a cardiac arrest, it, along with cardiopulmonary resuscitation (CPR), airway management and certain drugs, is the only options for attempting to avoid a patient's death. We normally take great care to avoid any encounter with electricity when working with electrical equipment, so it is surprising that at a point when our life is at most risk during a heart attack (cardiac arrest), we should turn to high voltage/high current to save the patient and use electricity to reset the heart back to a normal cardiac rhythm in order to reestablish blood circulation around the body. One point that always should be kept in mind is that a shock from a defibrillator does not 'start' the heart. What it does is to 'stop' the chaotic non-pulsatile cardiac arrhythmia such as ventricular fibrillation (VF), allowing for a short pause followed hopefully by the reestablishment within the heart of sinus rhythm.

DOI: 10.1201/9781003609414-3

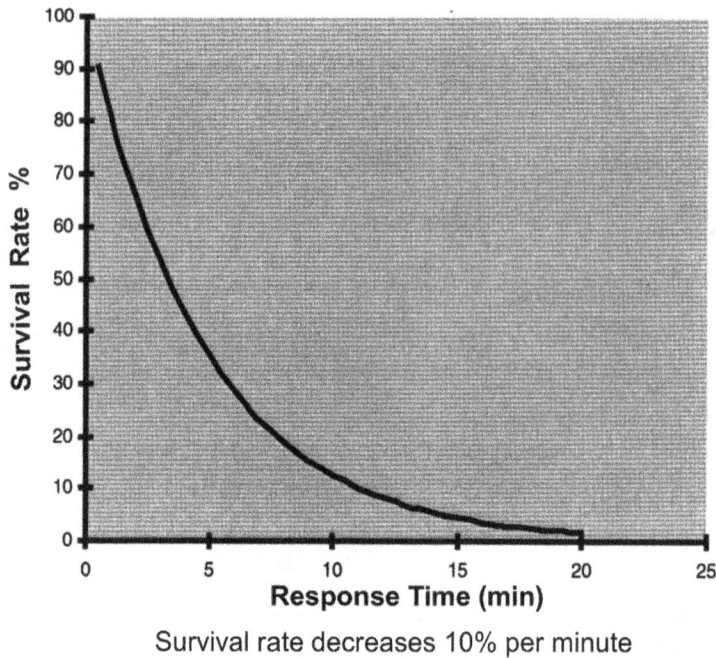

Survival rate decreases 10% per minute

FIGURE 3.1   Concept of early defibrillation.

Figure 3.1 shows that the use of a defibrillator is by no means a guarantee of the patient's survival. Time to defibrillation from the onset of a pulseless rhythm is very much a deciding factor and as can be seen in Figure 3.1. With each minute that passes before the shock from the defibrillator, the chances of the patient surviving reduce by 10% (UK National Library of Medicine). Due to lack of oxygenated blood flowing around the body because of the pulseless cardiac rhythm, organs such as the brain, kidneys and liver become impaired and damaged very quickly within minutes. The brain is particularly susceptible to damage within the first 5~6 minutes. However, if effective CPR be administered within the first few minutes, then the period of time to the first defibrillator shock can be extended. Although defibrillators are central to the process, they are just one of the steps in the 'chain of survival'

EARLY DEFIBRILLATION

Early defibrillation through a first-responder defibrillation program is the most important link to improve survival for adult cardiac arrest.

Studies have proved that the survival rate decreases 10% per minute. The delay to the time of the first defibrillator shock is the main determinant of survival in cardiac arrest, for both pre-hospital and in-hospital emergency resuscitation.

Automatic external defibrillators (AEDs) and **shock advisory defibrillators** have made defibrillation available to a much wider range of medical, nursing, paramedical and lay people. They have been specifically designed to meet the needs of emergency medical services (EMS) and in-hospital resuscitation teams.

Standardized training programs under medical control are to be established for these various groups.

The American Heart Association (AHA) and the European Resuscitation Council (ERC) have now included the use of AEDs in their resuscitation guidelines.

Heart Diseases

The number 1 killer

Cardiac Arrest
- The cause of approx. 12 % of
  all deaths in developed countries/year

Cardiac Arrest Events
- Min. 1000 events /Mil. Citizens /Year   (Europe)
- 2/3 of all Cardiac Arrest's happens pre-hospital
- 1/3 of all Cardiac Arrest's happens in-hospital

FIGURE 3.2

Before going any further, it is worth considering why defibrillators are now so readily available in almost every area where people gather. The numbers given in Figure 3.2 are very much a generalization that conveys approximately what the current situation is in many developed countries. Less developed counties will show figures that are radically different. As can be seen, cardiac arrests account for a sizable number of premature deaths around the world today.

Sudden Cardiac Arrest

STOP

- Definition
  - Sudden death due to onset of a
    sudden, chaotic and
    unproductive heart rhythm
- Events
  - 70-90 % Ventricular Fibrillation
  - 30-40 % Heart Infarcts
- Most effective VF Treatment
  - Defibrillation - *as early as possible*

FIGURE 3.3

A sudden cardiac arrest (Figure 3.3) is defined as a sudden chaotic cardiac electrical rhythm resulting in no or insignificant blood flow from the heart that would normally circulate around body (systemic system) and the lungs (pulmonary system). The majority of these non-productive rhythms are ventricular fibrillation (VF), but a significant number are pulseless ventricular tachycardia (VT) and asystole. A more detailed explanation of these rhythms will follow in this section.

The Solution to Sudden Cardiac Arrest

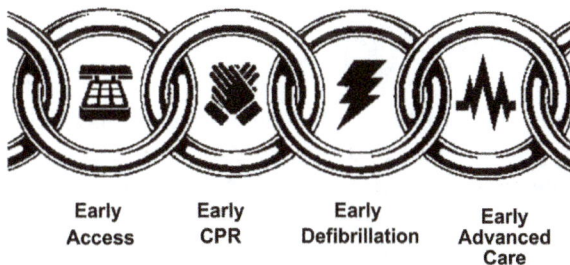

| Early Access | Early CPR | Early Defibrillation | Early Advanced Care |

FIGURE 3.4   The chain of survival.

Figure 3.4 shows the importance of a sequential approach in the event of a cardiac arrest. The elements within each step may vary from country to country, but essentially the basic steps are accepted worldwide. *'Early access'* refers to the need to call for professional help as the first step. Knowing the contact details to alert medical professionals such as paramedics, nurses and doctors is vital.

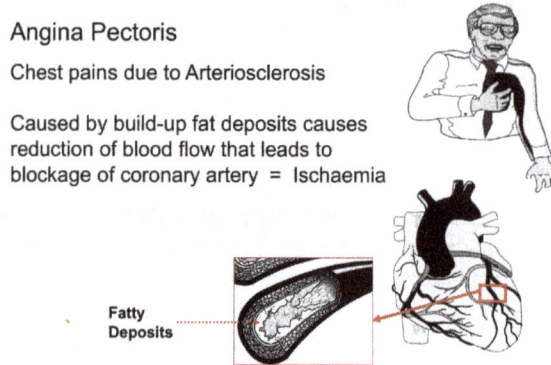

Angina Pectoris

Chest pains due to Arteriosclerosis

Caused by build-up fat deposits causes reduction of blood flow that leads to blockage of coronary artery = Ischaemia

Fatty Deposits

FIGURE 3.5   What can lead to a cardiac arrest?

The causes of coronary artery disease are many and varied. Smoking, diet, being overweight, hypertension (high blood pressure), high cholesterol (fatty deposits carried in the blood) and previous family history of heart disease all play a part in its development. Angina pectoris, chest pains due to arteriosclerosis, are often a very early sign that a patient may go on to have a cardiac arrest due to the buildup of fatty deposits in the coronary arteries (Figure 3.5). This is often referred to as ischemia (narrowing of the arteries), which results in inadequate blood supply (perfusion) to the heart muscle: the myocardium. During an episode of angina, the patient may feel severe chest pain and other symptoms such as breathlessness.

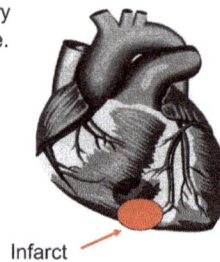

Acute Myocardial Infarction (AMI)

Serious blockage in coronary artery Infarct (death) of the cardiac tissue. Heart tissue does not regenerate AMI's can cause cardiac arrest First symptoms of AMI often overlooked

Infarct

FIGURE 3.6   What can lead to a cardiac arrest?

If the fatty deposits within the coronary artery continue to grow, they will eventually completely block the supply of oxygenated blood to the myocardium (the muscle layer of heart tissue) and within a few minutes this tissue becomes damaged and eventually dies (Figure 3.6). When this happens, the patient is said to have had a myocardial infarction (MI). This tissue does not recover and is problematic as it no longer responds to the electrical stimulus of the ECG signal or contracts. If the MI is sufficiently large, then the lack of contractibility of the myocardium will often lead to the patient's death.

1.	Sinus Rhythm	
2.	Tachycardia	

3.	Ventricular Tachycardia (VT), with pulse, below ~ 140/bmin	
4.	Ventricular Tachycardia without pulse, above ~ 140/bmin	

5.	Ventricular Fibrillation (VF), From course VF (around ~1.0mV amplitude) leading to fine VF. (Less than ~0.2mV amplitude)	

6.	Asystole	

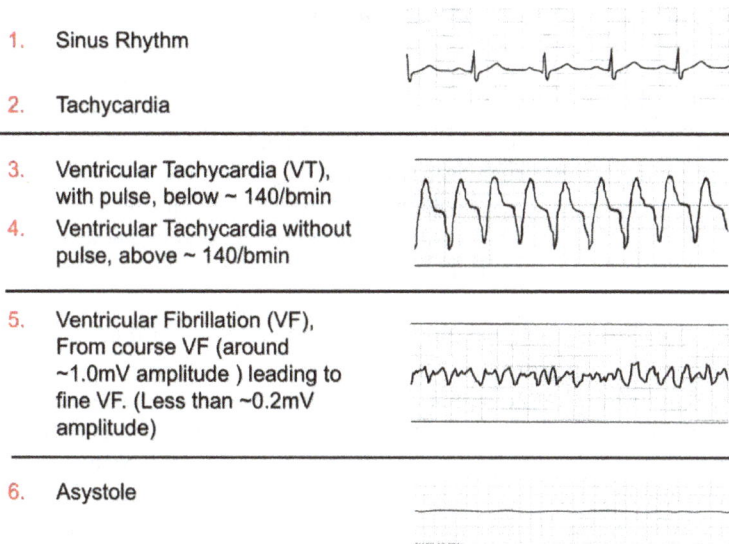

FIGURE 3.7   Possible sequence to MI and/or asystole.

Let us now look at a possible series of events and cardiac rhythms that lead to an MI and or asystole. This sequence above is by no means the only sequence that can lead to an MI and asystole.

In Figure 3.7, (**1**) and (**2**) are of sinus rhythm and tachycardia. The patient at this point is in normal sinus rhythm, perhaps sitting and relaxed. They then may attempt to climb a large flight of stairs, and as they climb, they become breathless, and their heart rate (HR) rises above 100beats/min. Although still in sinus rhythm, this rate is now tachycardia (adult HR above 100beats/min) as the body responds to the need to increase the oxygenation to muscles and the heart itself. When the heart is asked to maintain this excessively high HR, one important consequence is that the cardiac output, the amount of blood the heart circulates, begins to fall as the heart muscle is not given adequate time to fully refill in the diastolic phase before being entering the systolic phase.

(**3**) and (**4**) are VT. After a period of time in tachycardia, the coronary arteries and veins supplying the myocardium do not supply sufficient oxygen and fail to remove adequate carbon dioxide. This quickly leads to an increase in acidosis and a chemical imbalance between sodium and potassium within the myocardium muscle cells. The consequence of this is that a small group of cells within the ventricles begin to initiate, independently of the normal conduction signal, contractions within the ventricles and effectively become the 'pacemaker' of the heart. This ECG signal is very large and distinctive and a rate of below ~140/100beats/min will provide a reasonable amount of cardiac output but generally not enough to maintain the body's needs. Pulse VT like, sinus rhythm is classed as a 'none-shockable' rhythm and AEDs will not allow the operator to undertake to shock a patient.

After a short period of time the rate begins to climb, and the cardiac output again falls to the point where the patient may lose consciousness. At this point there is effectively no cardiac output, and the rhythm is classed as 'pulseless VT', a shockable rhythm. From this point, with the patient unconscious, the defibrillator may be used in order to stop the heart's chaotic electrical rhythm and re-establish normal sinus rhythm. Prior to defibrillation, basic life support, particularly CPR, will be needed, unless it is a witnessed cardiac arrest and defibrillation is available within the first minute, such as may happen in a cardiac ward.

(**5**) is VF. Left untreated after a few minutes the VT rhythm, either pulse or pulseless, may eventually convert to a VF form, in which there is no cardiac output, as small groups of cardiac cells within the myocardium begin to independently and spontaneously depolarize and repolarize. This chaotic rhythm would, if it could be seen through the chest wall, show the heart in what would appear in a tremor state with no contractions. Again, before a defibrillator is used, it is vital that CPR is undertaken in order to reestablish as much oxygenation and removal of carbon dioxide as possible in order to rebalance the sodium, potassium and pH levels within the myocardium conduction cells. Only when this has been achieved, is it possible to resynchronize all the cardiac conduction cells to one common state of polarization by means of the defibrillator's high current shock. Following a successful rebalance of the cardiac cells' chemistry and the establishment of polarization from the defibrillator shock, the heart is then given the opportunity to restart and re-establish sinus rhythm and circulation.

(6) is asystole. If the attempts at resuscitation prove unsuccessful or if no resuscitation attempts be made, then eventually VF will inevitably diminish and the rhythm will convert to asystole, a rhythm where there is no cardiac output and a slightly wandering flat line is observed on the monitor. This flat line often shows occasional P waves and even very intermittent ventricular activity but is now beyond the point where defibrillation shock will have any positive impact. It is classed on advisory defibrillators as a non-shockable rhythm and will not allow an operator to shock the patient. After approximately 15 minutes due to the lack of oxygen severe brain damage/death will occur.

Manual

- Traditionally used by emergency trained medical personnel in Critical Care Units, Intensive Care Units and Operating Rooms as well as by Paramedics in the pre-hospital setting. Unit may be used as an ECG bedside monitor.

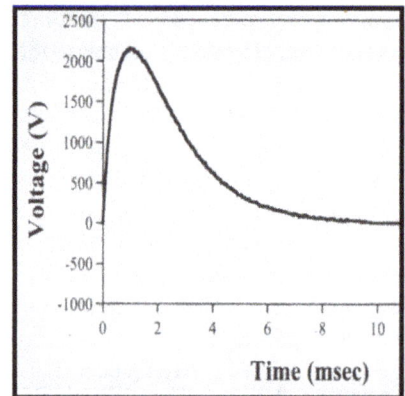

FIGURE 3.8   Defibrillator development.

Entirely manual defibrillators are now no longer available for purchase, but some are still in use around the world. These defibrillators were only operated by highly trained medical staff such as paramedics, doctors and nurses. To deliver the shock to the patient, the paddles were removed from their safety mounts and placed directly upon the patient's chest with either conductive gel pads or conductive gel, which was applied directly to the chest/paddles first. The operator would, without the assistance of software analysis, decide whether the rhythm was shockable and the amount of energy to be administered in joules. As you can see in Figure 3.8, the output was a mono-phasic DC pulse of electrical energy reaching over 2000 volts delivered over 4 m/sec.

Semi Automatic – Manual/AED

- Defibrillators/monitors that provide therapeutic and <u>diagnostic</u> functions. In the AED mode the Shock Advisory System advises the operator when it detects a shockable rhythm and requires user interaction in order to deliver a defibrillation shock.

- Traditionally used in the AED mode by Emergency Medical Technicians (EMT), Advanced Life Support (ALS), Basic Life Support (BLS) and other emergency medical personnel, who are trained to interpret arrhythmias and determine when a shock is needed.

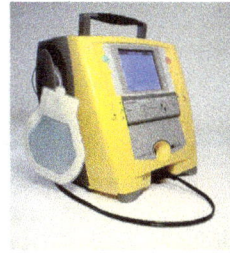

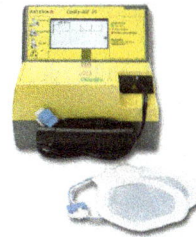

FIGURE 3.9 Defibrillator development – semi-automatic manual/AED.

Semi-automatic manual/AEDs (Figure 3.9) became available some 30 years ago and offered two advantages over the earlier defibs, in that they offered analysis of the cardiac rhythm and advised whether it was appropriate to shock the patient. It still requires the operator to charge the defib and administer the shock. There was also the ability with this design of defib to over-ride the analysis and operate the defib as a manual defib.

The second advantage was that the paddles were replaced with sticky pads that connected via an adaptor cable directly to the defib. This allowed the operator to shock the patient while at a safe distance in order to not make physical contact with the patient during the delivery of the shock. Around this time there was a change from the traditional mono-physic shock (DC) to a bi-physic shock (alternating current [AC]), which research has shown improved outcomes. Later in this chapter is full description of bi-phasic waveforms.

AED – Automatic External Defibrillator

- Lightweight, defibrillator units, fully controlled by microprocessors.

- A computerized voice instructs the operator to place two electrodes/pads on the patient's chest. The device analyses the heart rhythm and decides if a shock is necessary.

- The microprocessor will not permit the device to deliver a shock unless it detects the presence of a heart rhythm that requires defibrillation.

- These devices are used by emergency medical personnel, such as paramedics and nurses in less acute areas.

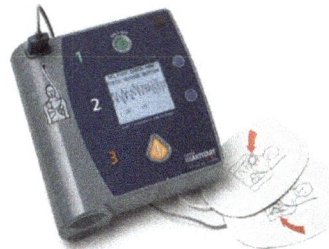

FIGURE 3.10 Defibrillator development – automatic external defibrillator (AED).

The advent of the portable small AED (Figure 3.10) has been the most significant step forward in saving lives in the past 30 years. There are several reasons for this, not least the significantly lower cost of the defibrillators. This enabled hospitals and other medical facilities to install these defibrillators in many wards and departments that had traditionally relied on the 'crash team', a team of medics that would urgently respond to a call via the hospital telephone switch board from a ward or department following a cardiac arrest. This team would be required to push a resuscitation trolley through the hospital in order to reach the ward or department where the cardiac arrest had occurred. This of course took valuable time and had a negative impact on the patient's chance of survival. With the placement of these small AEDs and appropriate department/ward staff training, the time to the patient receiving their first defibrillator shock was significantly reduced and outcomes improved.

With medical staff in these areas having less experience of using a defibrillator, the reliable software analysis ensures that only when a shockable rhythm is detected by the defibrillator would the defib charge and then instruct the user *'shock advised, stand back'*. It is a simple three step sequence.

Step 1: 'Apply pads'
Step 2: 'Analyzing rhythm'
Step 3: 'Shock advised, stand back' or 'No shock advised'

These defibs had a small display showing the patient's heart rhythm, and in some configurations would allow the operator to over-ride analysis software and operate the defib in manual mode.

■ PAD – Public Access Defibrillator

- Installed in public areas to reduce the time to perform defibrillation and improve the cardiac arrest survival rate.

- Most often these are 'Blind' defibrillators with no display

- Designed to meet the need for lay rescuer, who following training are able to perform BLS (Basic Life Support) and elements of ALS (Advance Life Support) including defibrillation

- Training courses enable the lay rescuers, including non-traditional first responders such as police, fire-fighters, airline personnel, security guards, senior care facility workers and other lay people to perform the essential CPR skills and to use a PAD.

- The unit should have the ability to electronically record information that can then be downloaded to a computer.

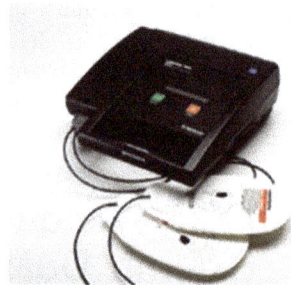

FIGURE 3.11   Defibrillator development–public access defibrillator (PAD).

The vast majority of cardiac arrests occur outside medical facilities, in public areas and homes. Therefore, in many countries it was decided that in order to significantly improve the survival rates for cardiac arrests, local and national programs of placing defibrillators and training the public in BLS and some elements of ALS such as defibrillation should be undertaken. Shopping malls, housing estates, sports grounds, swimming pools and stadiums, highway service stations and educational facilities such as universities are just some of the places that now have defibs available (Figure 3.11). Equally important to having defibs available is the need to have sufficient members of the public trained in CPR and the use of the defib, without this the effectiveness of these community defibrillation programs is greatly reduced.

Defibrillators designed for this specific use do not have a screen (blind) and are extremely simple to use. The vocal instructions guide the operator, a member of the public, with simple instructions such as 'continue CPR' and 'shock advised, stand back'. Minimal maintenance is required as the software within these defibs also undertakes regular checks of the electronic circuitry at predefined intervals such as weekly.

- Defibrillator protocols relates to the sequence and energy levels delivered to the patient during a cardiac arrest.

- There are several organisations and manufacturers around the world that decide these protocols

- The protocols are continually being developed in light of new research and are just one part of the overall 'Resuscitation protocol' adopted in each country and manufacturer.

- Always be aware of the defibrillation protocol adopted by your hospital/organisation and check that it is matched on any AED that you test or repair.

FIGURE 3.12   Defibrillation protocols.

As could be seen in the 'chain of survival', the overall protocol for resuscitation is a series of steps that should be followed in the event of a sudden cardiac arrest (Figure 3.12). These protocols are somewhat complex and detailed and differ from one country to another, and also may vary even within a particular country. Many western countries such as the United States, the United Kingdom, Australia and New Zealand have specific scientific committees such as the American Heart Association (AHA), Resuscitation Council UK and Australian and New Zealand Committee on Resuscitation (ANZCOR) that research and design the resuscitation protocols. These guidelines are always under constant review and regularly amended and updated. When this happens, organizations such as paramedic services and hospitals will undertake major retraining of staff to ensure that the new protocols are adhered to. The default sequence of energy levels is usually set by the manufacturer but can be varied to meet the requirement of individual counties or organizations where protocols vary.

- **Zoll - Rectilinear Biphasic**
  - Selectable, 3-shock sequence to accommodate escalating protocols. impedance compensation for constant current
  - Escalating from 120J - 200J (up to 360J)
  - "Constant Current Approach" in the first phase of the waveform.

- **Philips – Smart low energy Biphasic**
  - Biphasic Truncated Exponential (BTE) – Impedance Compensating
  - 3-shock sequence, non-escalating 150Joules,
  - Smaller 100uF capacitor (usual monophasic 200uF +)
  - Philips claim to have the largest number of peer-reviewed data of any defib manufacturer, both in-hospital and out-of-hospital

- **Medtronic – ADAPTIV™ Biphasic**
  - Energy up to 360J
  - Current Duration Adjustment & Voltage Adjustment
  - No need to change defibrillation protocol (200, 200, 360J)

FIGURE 3.13   Which Biphasic waveform?

Figure 3.13 illustrates three manufacturers' biphasic waveforms, demonstrating the principles behind this technology. While there are various manufacturers worldwide with their own unique approaches, biphasic waveforms represent a significant advancement over traditional monophasic waveforms.

Biphasic waveforms offer several advantages:

- *Higher Conversion Rates*: They have shown higher rates of converting atrial fibrillation (AF) and VT to sinus rhythm.
- *Lower Peak Voltage*: This ensures greater safety for both the patient and the operator.
- *Longer Shock Duration*: Biphasic shocks are delivered over a longer duration, typically up to 20 milliseconds, compared to the 3.2-millisecond duration of monophasic shocks.

Unlike monophasic waveforms, biphasic waveforms use AC, delivering a more complex shock pattern. This approach has been shown to improve effectiveness and reduce the risk of adverse side effects.

While there are variations among manufacturers' biphasic waveforms, the underlying principles and benefits remain consistent.

Each of the three different approaches relies on measuring the patient's chest impedance immediately before the delivery of the shock and/or at the period between the first and second phase. This allows the defibrillator software and electronic circuitry to adjust the current/voltage levels to ensure that the exact selected level of joules is delivered regardless of the patient's chest impedance. As can be seen in descriptions in Figures 3.14 and 3.15, the output waveforms and energy sequence may vary. Zoll's approach for the first phase is to ensure that it delivers a steady level of current and only vary in the delivery period time for the second phase—impedance compensation and ADAPTIV both.

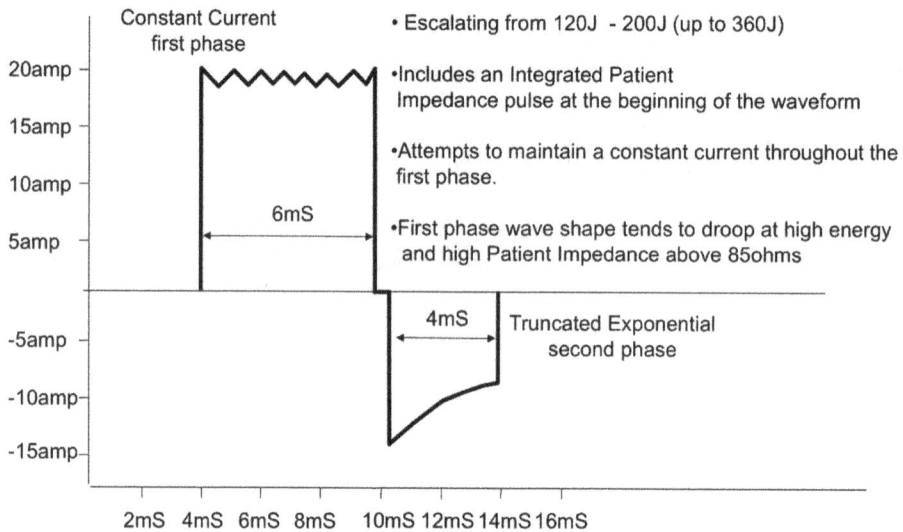

FIGURE 3.14   Zoll–Rectilinear Biphasic.

Zoll's approach to both the biphasic waveform and energy-level protocol is representative of the many differing ways a biphasic waveform and energy-level protocol can be designed. As can be seen in Figure 3.14, there is the ability if the operator has had advance operator training to deliver 360 Joules should they not achieve arrhythmia conversion to sinus rhythm using the standard escalating energy levels. The delay of ~4 milliseconds is to allow the electronic circuitry and software to measure the patient's chest impedance and set current/voltage in order to accurately deliver the set energy level.

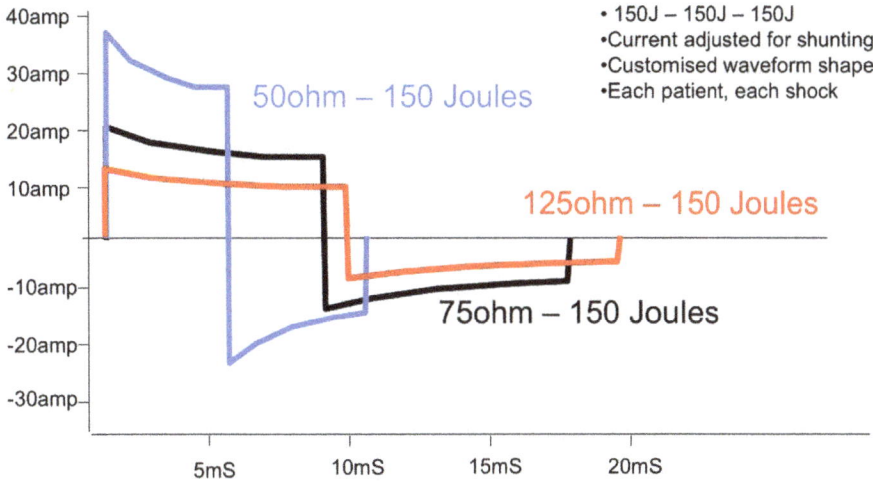

FIGURE 3.15 Philips – SMART biphasic impedance compensation.

Phillips' approach is again different in that each shock will vary the voltage, current and phase duration dependent on the chest impedance. Philips has proven with numerous clinical papers that a non-escalating energy level is highly effective in converting cardiac arrest patients to sinus rhythm.

FIGURE 3.16 Defib batteries.

As with all battery-operated medical devices, the batteries play a pivotal role in the device functioning (Figure 3.16). We all know the consequences of battery failure in such items as mobile phones and cars, so it is even more vital that we never underestimate the impact on patients should the battery within a medical device fail, for whatever reason. The electrical energy provided to the patient from a defibrillator can even be thought of as a life-saving drug. That may at first seem extreme but given its importance to the resuscitation of the patient, it is hard to avoid this analogy. All AEDs rely solely on their internal battery to operate and deliver shocks; other larger defibrillators used by paramedics, wards and departments also have internal batteries. In these larger defibrillators, if a close-by convenient mains outlet is available very near to the patient, then it is always advisable to plug the defibrillator into the mains supply, relying only on the batteries in the event of mains failure.

For a clinical engineer undertaking the service and repair of defibrillators, the testing and routine replacement of the defibrillator's battery is a crucial part of this work. All defibrillator manufactures will state when the battery pack should be replaced and also how and when testing of batteries should be undertaken. As with all equipment, records of the routine repairs should be kept accurately for future reference, particularly in the event of a defibrillator failure during a patient's cardiac arrest. Recording battery information such as batch numbers, serial numbers, stock received dates, battery cycling programs dates for rechargeable batteries and date fitted are just some of the information needed to be recorded in the service records.

Does battery analysis work? Yes, but it can only tell you the capacity of the battery on the day you undertook the analysis. It is not possible, from that information alone, to accurately predict when the capacity of the battery will fall to such a point that it is likely to fail during a resuscitation attempt of a patient. Be aware that it is not uncommon for the defibrillator to be required to provide more than ten shocks to a patient during a single cardiac arrest resuscitation.

4 Respiratory System and Mechanical Ventilation

THE RESPIRATORY SYSTEM

The Respiratory System

- The function of the respiratory system is to provide oxygen for the metabolic needs of the cells and to remove carbon dioxide, the cellular gas waste gas produced from metabolism

Like all mammals, humans breathe air and have lungs. The autonomic (involuntary) function of respiration is something that we all take for granted. Within moments of being born until we draw our last breath, it is central to our very existence. During an average lifetime of around 80 years, we might take around 670 million breaths. For an adult at rest, the normal respiratory rate is around 12–15 breaths per minute (bpm). The primary function of breathing is to provide oxygen and remove carbon dioxide.

As your experience with medical devices increases, two areas that you will undoubtably encounter are mechanical positive pressure ventilators and spirometry. Once you have a strong basic underpinning to your understanding of the anatomy and physiology of the respiratory system, you will be capable of being trained to undertake the repair and servicing of such devices as spirometers, continuous positive airway pressure (CPAP) machines and sophisticated mechanical patient ventilators. In some countries, such as the United States, some biomedical equipment technicians (BMETs) continue their careers by training further to become respiratory therapists, who are based in such departments as intensive care wards and respiratory therapy departments. In this role they are not only responsible for the repair and servicing of ventilators, but also prescribe the settings on oxygen therapy devices and some ventilators. They are very much medics and therefore work at the direction of doctors.

The diagrams in this chapter will portray in simple terms the anatomy and physiology of the respiratory system from the mouth, nose and upper airways all the way down to the alveoli and gas exchange. One very simple concept often used to describe the respiratory system is to think of it as an inverted tree. From the roots and base of the tree, (the mouth and nose), we then connect to the trunk of the tree (the trachea). The trunk then splits into large branches (the left and right bronchus), this then further divides in smaller branches (bronchioles). At the end of the small branches are even smaller stems that terminate in the tree leaves (alveoli sacs). A tree leaf absorbs carbon dioxide and releases oxygen, whereas in a human being, carbon dioxide is drawn from the returning blood supply and replaced with oxygen within the alveoli.

DOI: 10.1201/9781003609414-4

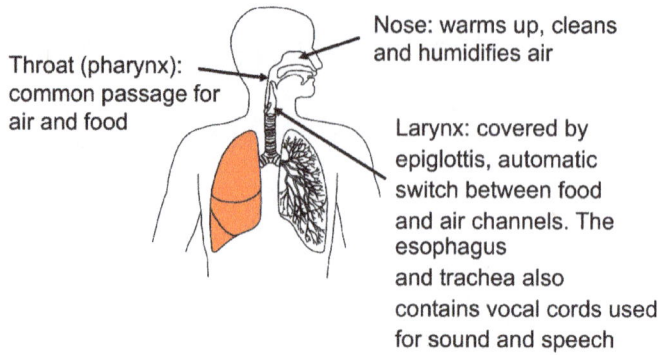

Nose: warms up, cleans and humidifies air

Throat (pharynx): common passage for air and food

Larynx: covered by epiglottis, automatic switch between food and air channels. The esophagus and trachea also contains vocal cords used for sound and speech

FIGURE 4.1.1   The upper airways.

The nose and mouth are the gateway to the rest of the respiratory system (Figure 4.1.1). Here, small particles such as pollen and dust are captured as they pass over small follicles in the nose and watery secretions (mucus) in both the nose and the mouth. Breathing, in the main, occurs through the nose, but the mouth is occasionally used when increased air flow is required, such as during rigorous exercise. An adult male at rest usually breaths a volume of around 500 mL/breath. Crucially, the resistance to air flow in both spontaneous and mechanical respiration, such as when a patient is connected to a mechanical ventilator, must be kept to a minimum. If resistance increases, the patient finds that they struggle to maintain an adequate volume of air for their needs.

The air entering the mouth and nose will be at temperature and humidity of the local environment. Both temperature and humidity can vary greatly. The lungs and the air within them "will" contain a great deal of water vapor at around 37°C, which is close to the core temperature of the body. This means that the upper airways and connecting vessels such as the trachea, carina and left and right bronchus must be capable of rapidly warming and humidifying the air intake on each breath. The atmosphere within the lungs is essentially tropical: warm and wet.

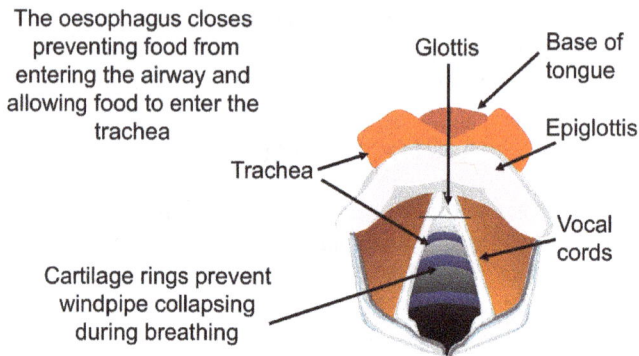

The oesophagus closes preventing food from entering the airway and allowing food to enter the trachea

Glottis

Base of tongue

Epiglottis

Trachea

Vocal cords

Cartilage rings prevent windpipe collapsing during breathing

FIGURE 4.1.2   Anatomy of the larynx.

The primary function of the larynx is to protect the trachea from fluids and food. One of the most amazing tissues in the human body is the epiglottis which sits above trachea and closes as fluids and food enter the upper airways. Its ability to react almost instantaneously ensures that neither fluid nor food enter the respiratory system. If this did happen, it would have a catastrophic impact on the patient's ability to breathe. Also, within this area, and shown in Figure 4.1.2, are the vocal cords, which vibrate to generate sound and speech.

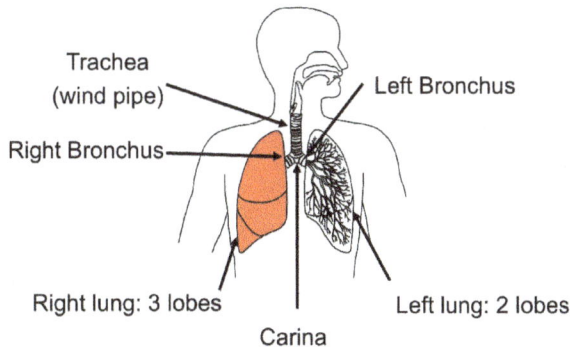

FIGURE 4.1.3 The middle airways.

The trachea, sometimes referred to as the 'wind pipe', is constructed with a series of 16–20 cartilage, 'C' rings, in a stack formation and connected by thin membranes between each ring. The flexibility of the cartilage allows for movement when exercising and it is approximately 15-cm long in an adult. At the base of the trachea is the carina where the trachea splits into the left and right bronchus. The left and right bronchus then enter into the lungs where they further divide (Figure 4.1.3). The two lungs are not identical as the right lungs has three lobes and the left lung has just two and is slightly smaller. This is due to the position of the heart which is slightly off-centered toward the left side of the chest.

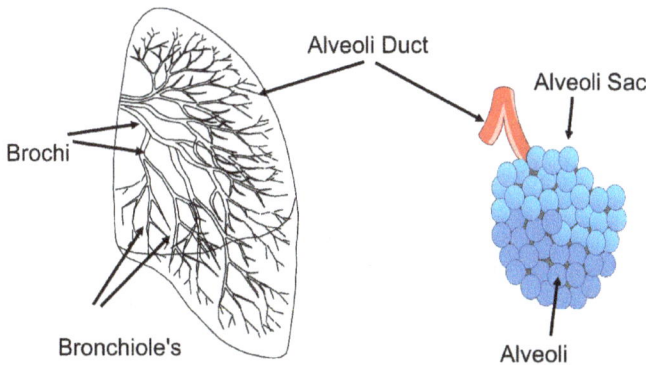

FIGURE 4.1.4 The lungs.

Within the lungs, the left and right bronchus divide further into bronchi, narrow airways that extend throughout each lung (Figure 4.1.4). These in turn again divide into smaller airways called bronchioles that again divide into alveoli ducts, the smallest of the airways. At the end of each alveoli duct is the alveoli sac that contains individual alveoli. An average person has approximately 500 million individual alveoli and that if it were possible to lay the alveoli flat on a surface would cover about the size of two tennis courts. This vast surface area is very important in the lungs ability to adequately allow for gas exchange of oxygen and carbon dioxide.

T = 40°C (104°F), RH = 100%
AH = 50mg/L

T = 25°C (77°F), RH = 45%
AH = 10mg/L

T = -15°C (5°F), RH = 7%
AH = 0.1mg/L

T = 20°C (68°F), RH = 50%
AH = 8.7mg/L

FIGURE 4.1.5   Different air conditions.

As stated above, humidity and temperature are crucial within the respiratory system. As you already know, the core temperature of the human body is given as 37°C or 98.6°F, so of course the temperature of the lungs will also be at the core temperature. What is vital to maintaining homeostasis, when the body is maintaining internal stability, is that air entering the lungs has a temperature very close to 37°C and maximum humidity. If the incoming air were cold and dry, then the body would lose heat and moisture as the air is expelled. This would rapidly lead to hypothermia (a core temperature below 35°C), with a dramatic deterioration in the patient's well-being.

Understanding the relationship between temperature and humidity is something that is central to understanding the respiratory system and mechanical ventilation. As the temperature of the air increases so does its ability to carry more water vapor. For an increase of each one degree Celsius, the quantity of water vapor a given volume of air can carry increases.

Relative humidity (RH): This is the maximum amount of water vapor at any given temperature the air can possibly hold. This is not to say that air at any given temperature is holding its maximum amount of water vapor, as it maybe that the air has not been in contact with sufficient water in order to collect the maximum amount of water vapor. What is given as the RH is a percentage figure for each degree Celsius, where 100% represents that the air is carrying the maximum amount of water vapor at a given temperature. If insufficient water is offered to the warm air, then the volume of water vapor held by the air will be lower, for example, 75%.

Absolute humidity (AH): This is another way of expressing the amount of water vapor, but this time in a given volume of air, 1 L. The amount of water vapor is expressed in milligrams per liter, for example, 10 mg/L. This expression of humidity is particularly useful in mechanical ventilation when administering active humidification to a ventilated patient.

For a ventilated patient, 'active' humidifier may be used. These medical devices come in a number of forms such as ultrasonic and evaporative and require the fresh gas flow to the patient to pass through them. There is also used a type of 'passive' in-line filter called an HME (heat and moisture exchanger) which is placed above the patient's connection to the endotracheal tube. As the patient breaths out, heat and moisture collect on the surface of the HME filter, so when the patient breaths in the heat and moisture from the filter then pass to the fresh gas.

If the air were to become cooler, then its ability to hold water vapor would also be reduced. When this happens to a ventilated patient, water droplets form in the tubing of the ventilator circuit. Clinicians often refer to this as 'rain-out'. Figure 4.1.5 shows a series of examples of the levels of humidity and temperature in various environmental locations.

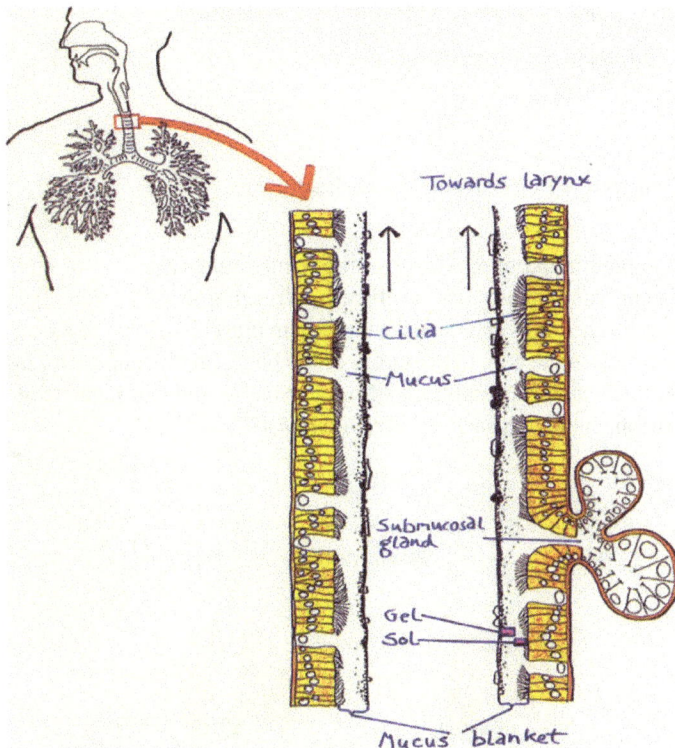

FIGURE 4.1.6 The mucociliary elevator.

The mucous and secretions add as much as 1000 mL of water per day to inspired air. The intimate contact between inspired air molecules and the nasal mucosa ensures that the RH of the inspired gas will be as high as 75%–85% by the time the air reaches the nasopharynx. The sub-mucus glands produce mucus gel that lines the trachea. This mucus lining captures inhaled debris and pathogens from the inspired air. This is a very important defense mechanism against viral and bacterial infections and debris such as pollen and dust. Once captured in the mucus, it is then transported back up through the trachea by the peristalsis action of the mucociliary elevator that then forces the flow of mucus up using the fine hairs of the cilia, eventually reaching the month (Figure 4.1.6). From here it can be swallowed and enter the digestion system where it is destroyed by acids in the stomach.

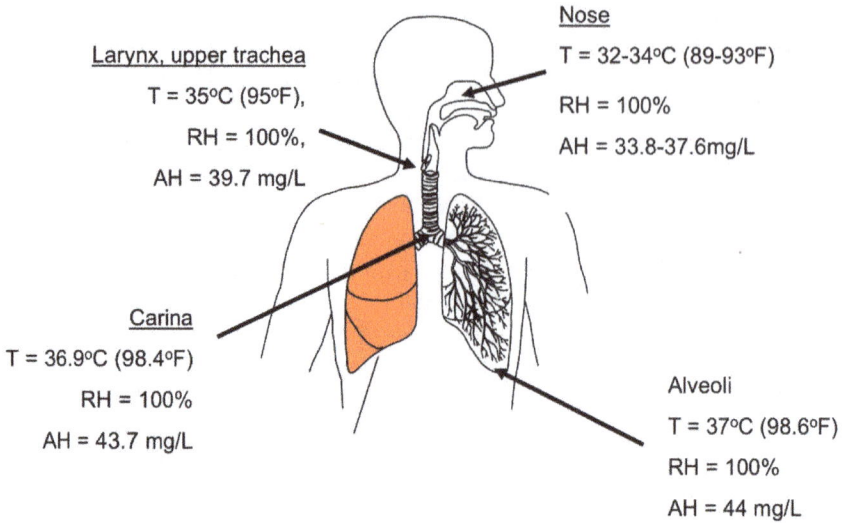

FIGURE 4.1.7 Humidity and temperature in the respiratory tract.

The air entering the mouth and nose will be at the temperature and level of humidity of the local environment. This can vary greatly from well below freezing to +40°C and have little or a large amount of water vapor. The lungs and the air within them must contain a great deal of water vapor at almost 37°C, which is close the core temperature of the body. So, the upper airways and connecting vessels such as the trachea, carina and left and right bronchus must be capable of rapidly warming and humidifying the air intake on each breath (Figure 4.1.7).

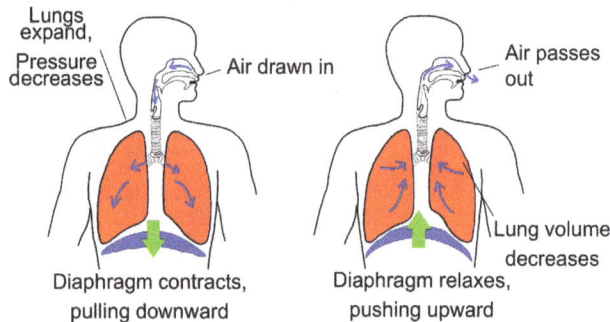

FIGURE 4.1.8 How respiratory system works?

A key distinction is that spontaneous ventilation is negative pressure ventilation, whereas mechanical ventilation is positive pressure ventilation where the machine drives air into the patient's lungs.

Having air move into and then out during respiration is achieved by the use of two sets of muscles. The first and largest of the muscles is the diaphragm which sits beneath the lungs on a horizontal plane between the abdomen and lungs. At rest it curves upwards and during inspiration it shortens, pulls down and straightens, thus creating negative pressure within the chest cavity. The second set of muscles involved is the intercostal muscles between each rib. At the start of inspiration these muscles expand which increases the size of the chest cavity which adds to the negative pressure within the chest cavity allowing the lungs to expand and thus draw in air (Figure 4.1.8).

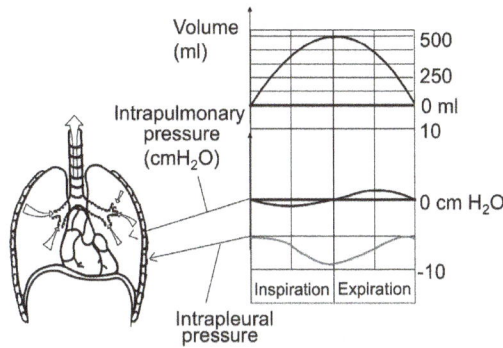

FIGURE 4.1.9 Mechanics of respiration.

There are several ways of expressing air pressure in respiratory medicine. The one used here is possibly the most common unit used and is the term 'centimeters of water' (cmH_2O). 1 cmH_2O is an extremely low pressure that would equate to the pressure you would feel on your cheek by a light breeze. You will also encounter kilopascals (kPa) and millibars (mbar) used to express air pressures.

Figure 4.1.9 shows the change in pressures and air volume within lungs and the intrapleural sac that surround the lungs for an average adult. At the start of the inspiration phase, the effects of the diaphragm contracting and intercostal muscles expanding generate a rapid drop in pressure in the intrapleural sac from about −5 cmH_2O to around −10 cmH_2O. This constant −5 cmH_2O negative pressure ensures that the small airways and alveoli sacs within the lungs remain open and do not collapse. The air at the start of inspiration is drawn into the lung as the atmospheric air is at a higher positive pressure than the lungs internal negative pressure. The pressure within the alveoli will be around −1 cmH_2O on inspiration and +1 cmH_2O on expiration. Due to the very low resistance to air flow by the upper and central airways, a surprisingly high volume of air (500ml) is able to reach the lungs. This volume of air is referred to as the Tidal Volume (Vt or TV) as it flows in then out with each breath as a tidal flow.

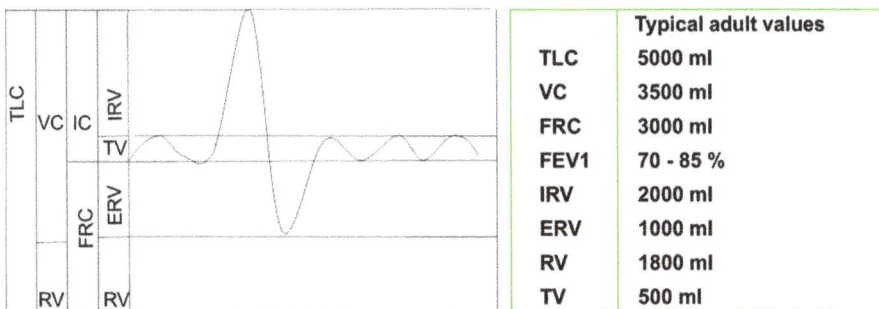

Typical adult values	
TLC	5000 ml
VC	3500 ml
FRC	3000 ml
FEV1	70 - 85 %
IRV	2000 ml
ERV	1000 ml
RV	1800 ml
TV	500 ml

FIGURE 4.1.10 Lung volumes and capacities. *Abbreviations*: ERV, Expiratory reserve volume; FEV1, Forced expiratory volume in the first second of expiration; FRC, Functional residual capacity; RV, Inspiratory reserve volume; RV, Residual volume; TLC, Total lung capacity; TV, Tidal volume; VC, Vital capacity.

In order to achieve oxygenation and the removal of carbon dioxide, the volume of air within the lungs must be sufficient. Following on from the previous, Figure 4.1.10 expands on various parameters that can be measured using a small hand-held device called a spirometer that is connected to a computer. The clinician, often a respiratory therapist, asks the patient to undertake various breathing patterns. The clinician can use the information gained, alongside other clinical information, to diagnose many respiratory diseases and conditions such as asthma or chronic obstructive pulmonary disease (COPD). What follows is a brief description of some of the terms used in spirometry. They relate to an average adult male's lung.

TV = Tidal Volume: Normal amount of air that passes in and out of the lungs while the patient is at rest.

TLC = Total Lung Capacity: This is the maximum volume the lungs can hold. It is composed of the following:

TLC = Residual Volume (RV) + Expiratory Reserve Volume (ERV) + Tidal Volume (TV) + Inspiratory Reserve Volume (IRV)

This value is not measured by a spirometer but is a calculation undertaken with other factors such as the patient's size and weight. The lungs are one of the largest organs in the human body, which can be better understood if you visualize their capacity in terms of liquid, for instance with 5 X 1-L bottles of water.

VC = Vital Capacity: This is the total amount of air exhaled after maximal inhalation and maximum exhalation. Composed of the following:

VC = Inspired Reserve Volume (IRV) + Tidal Volume (TV) + Expired Reserve Volume (ERV).

FRC = Functional Residual Capacity: The volume remaining in the lungs after a normal, exhalation. Composed of the following:

FRC = Expired Residual Volume (ERV) + Residual Volume (RV)

FEV1 = Forced Expiratory Volume in the first second of expiration: This percentage figure is extremely useful in determining resistive airways. A healthy individual would normally show a reading of over 80%. For this test, the patient is asked to breath in as much as they can and then breath out as hard and as fast as they can manage.

IRV = Inspiratory Reserve Volume: This is amount of air that can be forcibly inhaled after a normal tidal volume.

ERV = Expiratory Reserve Volume: The additional volume of air that can be expired with the patient's maximum effort beyond the end of a normal expiration.

RV = Residual Volume: The volume of air remaining in the lungs after maximum forced expiration.

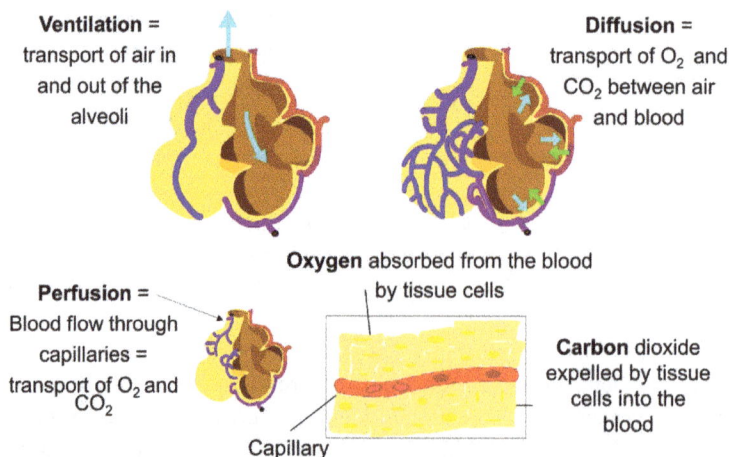

FIGURE 4.1.11 Gas exchange.

As described before, the final and most import components of the lungs are the alveoli and alveoli sacs (Figure 4.1.11). It is here that the exchange of oxygen from the air and carbon dioxide from venous blood return and pass in opposite directions. In order to effectively ensure this can happen, it is required that sufficient air reaches and inflate the alveoli (ventilation) and that there is an adequate supply of deoxygenated blood flowing around the external sides of the alveoli sacs (perfusion). With these two conditions met, it should result in gas exchange of oxygen and carbon dioxide (diffusion).

The bottom right-hand side of Figure 4.1.12 shows the diffusion happening in the other tissues and organs throughout the body. This of course is in the opposite directions to that in the alveoli but uses the same principal.

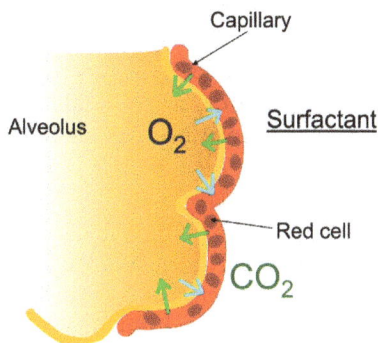

FIGURE 4.1.12   Pathways of O_2 and CO_2 diffusion.

Surfactant is a secretory product, composed of lipids (fatty, waxy or oily compounds) and proteins (complex molecules found in all living organisms) and is made within the lungs. It may be thought as a 'gas grease' that reduces surface tension within the alveoli sacs and allows for an easier exchange of oxygen and carbon dioxide through the wall of the alveoli. Without sufficient surfactant production breathing becomes much more difficult, which can lead to many severe conditions. Today we are able to treat this problem by the replacement or supplementing of exogenous surfactant (produced within a laboratory) to the alveoli as an aerosol or liquid down the endotracheal tube. The advent of this treatment has led to a major step forward in the treatment of premature neonates who do not begin to produce surfactant until about 26 weeks of pregnancy.

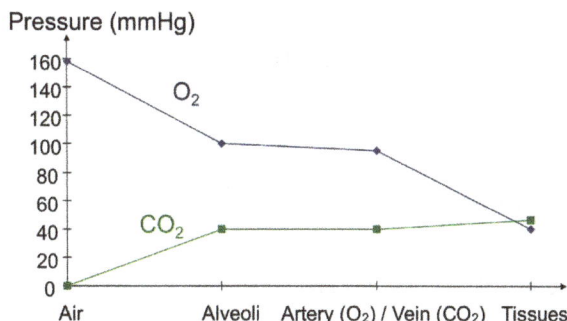

FIGURE 4.1.13   Partial pressures of gases.

The question you may now be asking is what drives the migration of oxygen and carbon dioxide in opposite directions within the alveoli, arteries, veins, tissues and organs? This will happen due to an imbalance of gas pressures across surfaces such as the alveoli walls. Higher concentrations naturally attempt to disperse toward lower adjoining concentrations.

WHAT IS A PARTIAL PRESSURE?

As air at sea level has a barometric pressure of 1 atmosphere, which is to say 760 mmHg, and oxygen makes up 1/5th of air, then it can be said that as a 'partial' component of the air, it therefore accounts for 1/5th of the total pressure, approximately 160 mmHg.

OXYGEN

If we look at Figure 4.1.13, we can see that the oxygen pressure within the capillaries around the alveoli is approximately 100 mmHg compared to the oxygen content of the air within the alveoli of 160 mmHg. This pressure difference of some 60 mmHg is what drives the migration of oxygen from the air into the alveoli capillaries. Continuing to follow the oxygen line to the right on the diagram, we next encounter arterial oxygen when it reaches tissues and organs. Here the pressure gradient is 100 mmHg to about 40 mmHg, a difference of again some 60 mmHg. This difference again drives the uptake of oxygen from arteries and arterioles into surrounding tissues and organs.

CARBON DIOXIDE

Starting at the right-hand side of the diagram, we can see the carbon dioxide is at its highest concentration within the tissues/organs, around 45 mmHg pressure. The level within the veins is around 40 mmHg, meaning that there is only a very small difference of just 5 mmHg. Migration from tissue/organs into the veins is greatly aided by the fact that carbon dioxide molecules are far more soluble in plasma (colorless fluid part of the blood) than oxygen molecules, so even with a lower driving pressure gradient of just 5 mmHg, it is sufficient to ensure the removal of most of the carbon dioxide from tissues/organs. From tissues/organs, the venous blood returns to the lungs where it encounters a far larger pressure gradient of some 40 mmHg. As air contains almost no carbon dioxide, migration from alveoli capillaries through the wall of the alveoli is relatively easy.

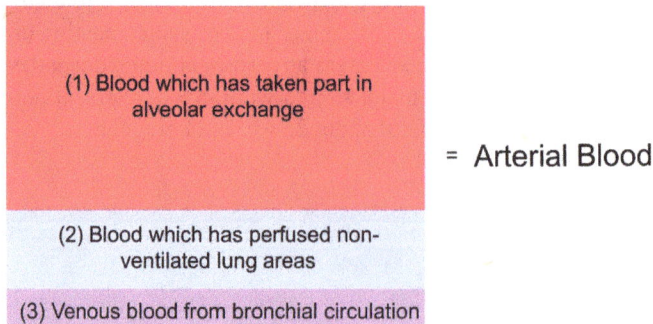

FIGURE 4.1.14   Pulmonary perfusion.

The three sections in Figure 4.1.14 show just how much of the pulmonary venous return is actively engaged in gas exchange (1) around the alveoli. During a restful state, such as sitting, only about 2/3rd of the pulmonary venous return is actively engaged in gas exchange. As breathing is steady but relatively shallow, lower parts of the lung's lobes (2) do not receive sufficient ventilation of air to provide any meaningful gas exchange. This state will change rapidly once the individual becomes active—standing, walking and undertakes vigorous exercise—thus utilizing these lower lobes during deeper breathing. As the lungs (3) are not only constructed of alveoli but also bronchus, bronchi and bronchioles, these must also receive their own blood supply which of course is not actively engaged in gas exchange.

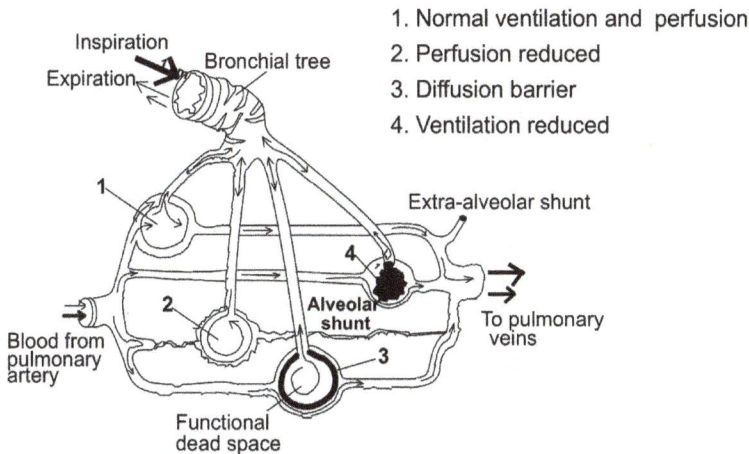

1. Normal ventilation and perfusion
2. Perfusion reduced
3. Diffusion barrier
4. Ventilation reduced

FIGURE 4.1.15   Problems with ventilation and pulmonary perfusion.

Figure 4.1.15 shows a representation of alveoli sacs and the perfusion around them. Take some time to study each of the four situations and the impact on gas exchange.

- *Normal ventilation and perfusion*: This is an ideal situation where there is a correct level of inflation within the alveoli sac and adequate perfusion around it, creating a condition for maximum gas exchange.
- *Perfusion reduced*: Here there is sufficient ventilation, but possibly due to disease, insufficient perfusion, which will greatly reduce gas exchange. If during mechanical ventilation the air pressure within the alveoli is excessive, it will exert pressure on the capillaries reducing blood flow.
- *Diffusion barriers*: This can have several reasons, one of the most common is a thickening of the alveoli wall, which will impede the ease of gas exchange. Other medical conditions such as pulmonary edema (a build-up of fluid within the lungs) and pulmonary fibrosis (a scarring within the lungs) also greatly inhibit gas exchange.
- *Ventilation reduced*: A number of reasons can impact on the level of ventilation within the alveoli sacs. Narrowing of the airways due to conditions such as asthma and chronic bronchitis are two of main reasons, insufficient airway pressure and flows during mechanical ventilation are another reason.

Also mentioned above are alveoli shunts, where perfusion may not have been in contact with the surrounding surfaces of the alveoli sacs and therefore not available for gas exchange.

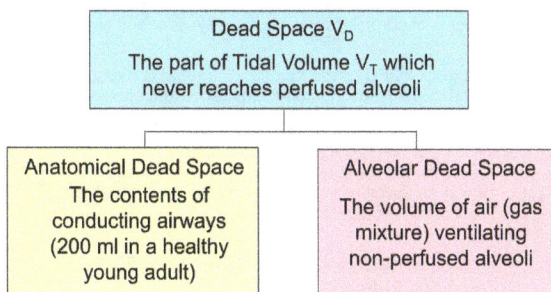

FIGURE 4.1.16   Dead space.

Dead space is air that is inspired but does not actively engage in gas exchange in the alveoli. It can be thought as the air volume that supports the air within the alveoli. Figure 4.1.16 relates not only to spontaneous breathing but also mechanical ventilation. In mechanical ventilation when setting a tidal volume for a patient, the clinician will also add additional volume to account for dead space.

Anatomical dead space: If we were to say that a typical adult lung during normal, at rest, breath had a typical alveoli volume of 500 ml of air, then in order to deliver this volume to the alveoli sacs, we would also need to provide an additional 200 mL of air to fill the proceeding airways such as the trachea and bronchi and bronchioles.

Alveolar dead space: As described in Figure 4.1.16 not all of the alveoli sacs are supplied with sufficient perfusion to allow for gas exchange to take place. Pulmonary diseases such as emphysema can have a large impact on the number of alveoli sacs available for gas exchange.

One other form of dead space you may encounter is called mechanical dead space. When a sedated patient is prepared for surgery, then an endotracheal tube (ET Tube) is inserted down their throat and terminates at the mouth. Very often an additional connector is then connected and effectively extends the ET tube and provides small connector ports to allow for gas/air to be drawn away and taken to a patient monitor for gas analysis and also the measurement of pressures and volumes (spirometry). This addition connector also requires additional volume to be taken into account and has the additional required volume printed on the side of connector of approximately 40 ml.

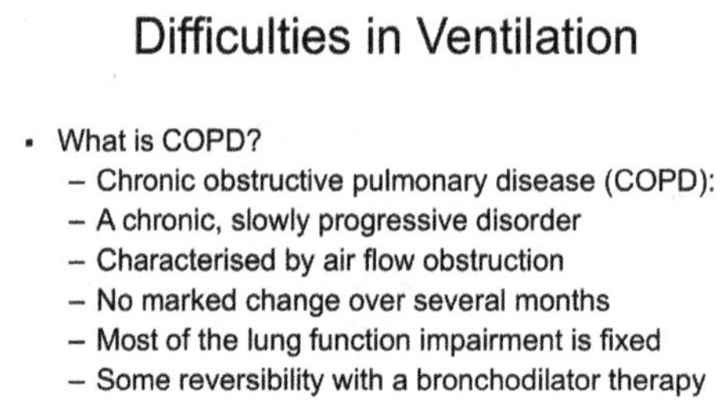

Difficulties in Ventilation

- What is COPD?
  - Chronic obstructive pulmonary disease (COPD):
  - A chronic, slowly progressive disorder
  - Characterised by air flow obstruction
  - No marked change over several months
  - Most of the lung function impairment is fixed
  - Some reversibility with a bronchodilator therapy

FIGURE 4.1.17

The two most common respiratory diseases are COPD (Figure 4.1.17) and asthma. Both are classed as chronic diseases meaning that they last more than one year and require regular medical attention. They also negatively impact on an individual's normal daily life by limiting physical activity. COPD occurs in adult life and requires treatment through means such as bronchodilators, including fluticasone (Flovent) which comes as an inhaler used twice daily. Budesonide (Pulmicort) is available as a handheld inhaler or for use in a nebulizer. Prednisolone comes as a pill, liquid or injection. Great strides are currently being made in the treatment of COPD, but it is unlikely that there will ever be a cure.

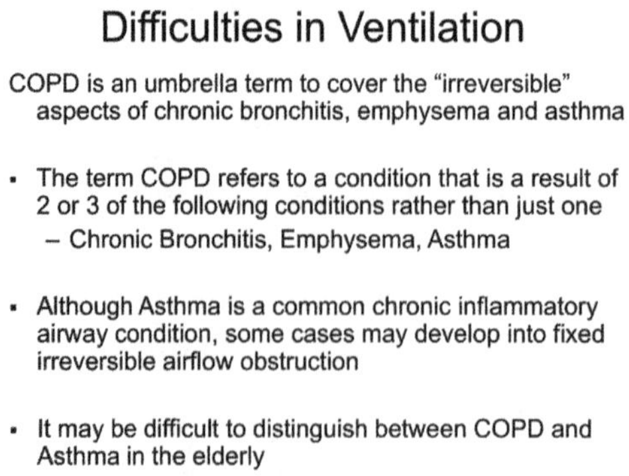

Difficulties in Ventilation

COPD is an umbrella term to cover the "irreversible" aspects of chronic bronchitis, emphysema and asthma

- The term COPD refers to a condition that is a result of 2 or 3 of the following conditions rather than just one
  - Chronic Bronchitis, Emphysema, Asthma

- Although Asthma is a common chronic inflammatory airway condition, some cases may develop into fixed irreversible airflow obstruction

- It may be difficult to distinguish between COPD and Asthma in the elderly

FIGURE 4.1.18

As can be seen from Figure 4.1.18, COPD is an umbrella term that covers two or three conditions that a patient may be suffering from at a given time.

Asthma is a condition that occurs as a reaction to airborne particulates such as pollen and dust. As a result of these, other trigger particulates cause the airways to narrow and swell and may produce extra mucus leading to a shortness of breath. It can be so severe that it leads to respiratory failure and even death.

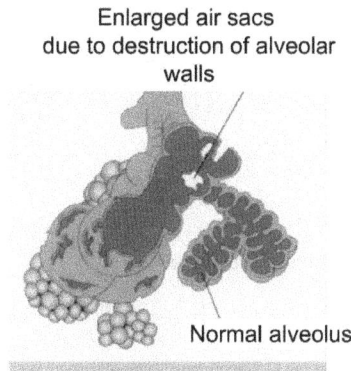

Emphysema

• Abnormal, permanent enlargement of the alveolar air spaces

• Shortness of breath is a classic symptom

Enlarged air sacs due to destruction of alveolar walls

Normal alveolus

FIGURE 4.1.19 Difficulties in ventilation.

Emphysema is generally caused by cigarette smoking or long-term exposure to certain industrial pollutants and environmental factors such as pollution and dust. The destruction of the alveoli sacs results in far less surface availability for the gas exchange of oxygen and carbon dioxide (Figure 4.1.19).

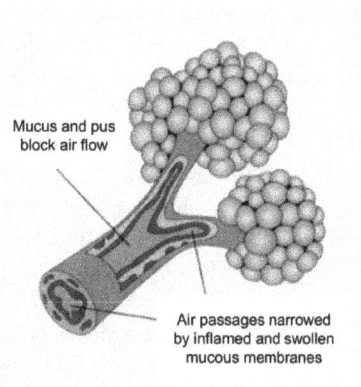

• Chronic Bronchitis
  – Presence of a chronic cough and sputum production
  – Chronic cough occurs for at least 3 months in 2 consecutive years
  – Absence of other recognised causes of sputum production

Mucus and pus block air flow

Air passages narrowed by inflamed and swollen mucous membranes

FIGURE 4.1.20 Difficulties in ventilation.

Chronic bronchitis is the result of an over production of sputum in the bronchus, bronchi and bronchioles and results in effectively narrowing the airway passages (Figure 4.1.20). Again, the most common reason for bronchitis is smoking. The most common symptoms are as follows: frequent coughing or a cough that produces excessive mucus, sometimes referred to as 'smokers cough'; wheezing, a whistling or squeaky sound when you breathe; shortness of breath, especially with physical activity; and tightness in the chest. Clinicians will use a stethoscope to listen to a patient's chest in order to hear wheezing or percolating sounds in the lungs when diagnosing bronchitis.

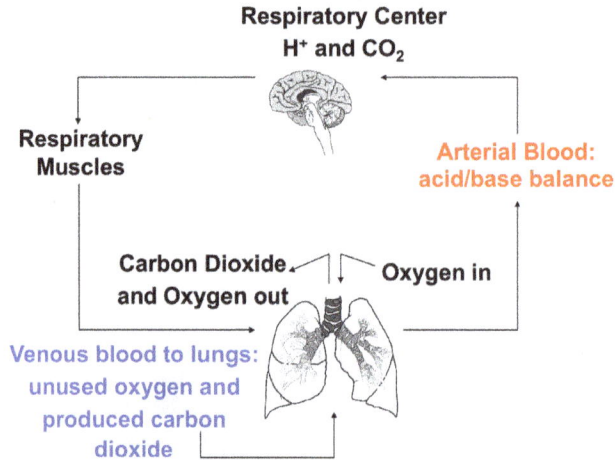

FIGURE 4.1.21   Regulation of breathing.

The power of hydrogen (pH) is a measurement used to define the acidity or alkalinity of solutions such as the blood. In the human body, we normally have a pH reading between 7.35 and 7.45 with an average of 7.40. Hydrogen ions play the crucial part in determining the bloods pH and are very much affected by the level of carbon dioxide in the blood stream. With excessive carbon dioxide in the blood stream, we will see a fall in pH, this can have severe detrimental impacts on organs such as the heart, brain and liver. A pH level of below 6.9 will eventually lead to coma.

The control of breathing is undertaken at the base of the brain in an area called the medulla oblongata (medulla). It is here that the monitoring of various levels within the blood stream is undertaken such as carbon dioxide, oxygen and pH levels. The regulation of breathing involves the control of the rate of breaths and the depth of the breaths (Figure 4.1.21). It is vital that volume and rate of breathing is the correct level to control both the level of oxygen and carbon dioxide in the blood stream. Hyperventilation (abnormally high rate and depth of breathing) can deplete the level of carbon dioxide in the blood stream and thus change the blood pH to become alkalotic to the point that the patient may lose consciousness and passes out if spontaneously breathing. Hyperventilation often occurs when an individual encounters severe stress or sudden shock which causes them to hyperventilate. Hypoventilation (low respiratory rate and depth of breath) is equally problematic causing a drop in the pH level and a rise in acidity in the blood.

If patients were to have their inspired and expired gasses measured by means of a side-stream gas sample line connected between a endotracheal tube and patient monitor, then typically we would see values on inspiration of 21% oxygen (FiO_2, fractional inspired oxygen) and 0% carbon dioxide ($FiCO_2$, fractional inspired carbon dioxide). On the expiration phase then typical values of 17% oxygen (FeO_2, fractional expired oxygen) and 5% carbon dioxide ($FeCO_2$, fractional expired carbon dioxide) are to be expected. Note, that the percentage amount of carbon dioxide expired will not necessarily equal the amount of oxygen absorbed. This is very much determined by the level of metabolic activity of the patient. Metabolic activity is the bodies' consumption of fats, proteins and carbohydrates within the various organs of the body, which produce heat, growth, energy, cell replacement and fights infections.

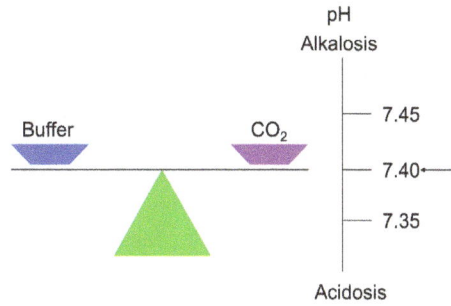

FIGURE 4.1.22   Acid-base balance.

As the pH of the blood is very dependent on the patient's level of ventilation, then in order to mini-mize the swings in pH the body uses a chemical offset system (buffer) to continually adjust the pH to stay around 7.40 (Figure 4.1.22). Please be aware that organs and systems throughout the body also regulate their own pH levels which are often different from that within the blood stream.

SECTION 4.2

MECHANICAL VENTILATION

As set out in the foreword, this textbook is designed for clinical engineers at the start of their careers with little or no experience of medical devices. As mechanical ventilators are highly sophisticated critical devices, they require formal training from the ventilator manufacturer, their agents or from within your organization. It is important that training records are kept, and competencies regularly reviewed. What is offered here is only an overview of the basic principles. Only after a number of years of clinical engineering experience will a clinical engineer be selected for training and certifi-cation in order to work on patient ventilators.

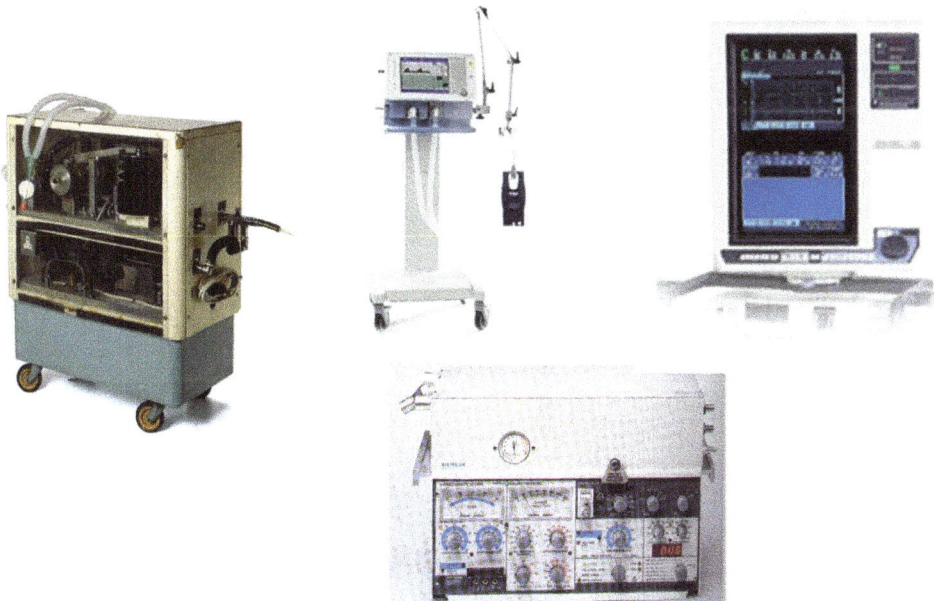

FIGURE 4.2.1   Mechanical positive pressure ventilation.

In order to gain most of the knowledge from this chapter, one should have completed the section on the respiratory system, as many of the terms and functions of a mechanical positive pressure ventilator are directly related to the physiology of the respiratory system. At this point, it is worth mentioning that hospitals around the world often use different titles for departments that undertake similar patient care work. It is therefore worth having an understanding of the interchangeable department names The term used in the United States for the emergency room is emergency room (ER), whereas in the United Kingdom, this can be the accident and emergency department (A&E). The intensive care unit (ICU) can also be known as the critical care unit (CCU) or intensive therapy unit (ITU). The room where surgery is performed is called the operating room (OR) in the United States, but in the United Kingdom it is referred to as the operating theater (OT). There are also other department titles that can be used for pediatrics such as pediatric intensive care unit (PICU) and neonatal intensive care units (NICU), which can also be called special care baby units (SCBU).

This chapter is only a brief introduction to the complex subject of mechanical ventilation and not a definitive full and comprehensive discussion. One of the most important things to understand when discussing mechanical positive pressure ventilation is that the machine is only involved in driving air into the patient during the inspiratory phase only, and not drawing air out of the patient during the expiratory phase, which is solely driven by the patient's diaphragm and inter-costal muscles. Modern modes of ventilation can assist the patient to breathe out, but do not actually draw air from the patient. A basic principle that is almost always adhered to is that the required volume should be delivered to the patient for the minimal amount of pressure required to do so.

One of the most important advancements in medical technology has been the development of today's mechanical positive pressure ventilators (Figure 4.2.1). The ability to ensure that a patient remains ventilated even after suffering traumatic injury, infection or the effect of anesthesia means that a patient's survival chances are dramatically improved. The early ventilators originated in the early 1950s and were commonly called 'Respirators'. They often worked on a piston drive system or bellows that would drive a given volume of air into the patient via an 'un-cuffed endotracheal tube'. Manufactures such as Drager, Bennett (now Puritan and Bennett) and Cape Ventilators were early pioneers and some of these companies are still manufacturing ventilators today.

Two very important terms that should always be remembered when talking about mechanical ventilation:

1. *'Invasive'* delivery, it is when an endotracheal tube is placed down the patient's throat and into the trachea.
2. *'Non-invasive'* delivery, it is where a sealed face mask is placed around the patient's nose and mouth.

There are also other methods such as a tracheostomy where an opening is made in the neck and into the trachea, allowing the insertion of a tracheostomy tube by which the patient is able to breathe spontaneously or be connected to a mechanical ventilator.

The two reasons and three areas for Mechanical Ventilation

- Emergency hospital/field settings (i.e. road traffic accidents). **Maintenance.** Bag, valve and mask operated by paramedics/doctors in order to transport patients

- Operation Room. **Maintenance** of patient oxygenation and delivery of anesthetic agents whilst the surgeon operates. Mechanical Ventilators

- ITU/CCU. **Therapy,** used to actively improve the paralysed patient's condition, both respiratory and cardio vascular. Advanced Mechanical Ventilators.

FIGURE 4.2.2

Used in emergency hospital and field settings, the bag, valve and mask (often referred to as an 'AMBU bag') are vital tools that can be rapidly deployed in order to ventilate a patient whose breathing has either completely ceased or is insufficient to maintain oxygenation. The major advantage in its use is that there is no need to intubate the patient, which allows for rapid deployment in an emergency situation. Patients can sometimes be electively 'bagged' when being transported between the OR and the CCU (Figure 4.2.2).

Within the OR, the OR ventilator is integrated within the anesthesia machine. In the early days of OR ventilation, these ventilators were simplistic in their operation, and were relied upon to maintain the patient's basic respiratory needs during surgery. However, as mechanical ventilation has developed many of the more advanced intensive therapy unit's (ITU) ventilator feature, modes of ventilation have now been integrated into the OR ventilator. The benefit of this is that a more individually tailored ventilation mode and settings can be used, which results in less stress to the respiratory system and improved recovery outcomes.

The CCU's ventilators are some of the most sophisticated devices found in a modern hospital. They can support the totally respiratory paralyzed patient, who is totally dependent on the mechanical ventilator, all the way through to the patient breathing unaided and spontaneously for themselves. These highly complex machines are able to work in synchrony with the patient as they recover, using a number of ventilation modes that are specifically designed for each stage of ventilation recovery. Providing the correct and tailored support at each stage of the recovery process also has a positive impact on other physiological systems such as the cardiovascular system. It can therefore be thought as a therapeutic device.

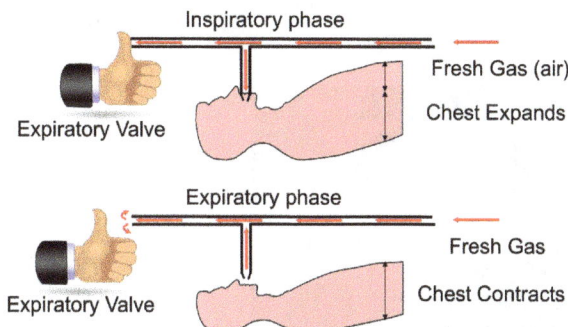

FIGURE 4.2.3 Simple positive pressure-controlled ventilation.

At its very simplest, positive pressure ventilation is achieved by blocking the escaping fresh gas (air) at the end of the 'T' circuit which then forces the flow into the patient's lungs causing the chest to rise. Once a given pressure or volume has been delivered to the lungs, the expiratory valve opens, and due to the patient's diaphragm and intercoastal muscles working together, the air within the lungs is then expelled. No matter how complex a positive pressure ventilator may appear, they all operate using an expiratory valve (Figure 4.2.3).

You may also encounter ventilators that also have an inspiratory valve. This is placed in line with the fresh gas input and blocks the flow of fresh gas once the required pressure or volume is achieved. At this point both expiratory and inspiratory valves are closed thus holding the air within the patient's chest for a required period of time. This is referred to as inspiratory pause.

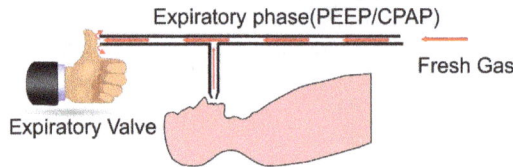

FIGURE 4.2.4   Simple positive pressure ventilation.

Two of the common terms used in positive pressure ventilation are positive end expiratory pressure (PEEP) and continuous positive airway pressure (CPAP). In both modes a small amount of back pressure is maintained by not fully opening the expiratory valve to allow all the air to escape (Figure 4.2.4) This is usually a very low pressure, in the region of 2–5 cmH$_2$O. This has the effect of preventing the collapse of the small airways and alveoli sacs during expiration. The reason for the two separate terms is to denote whether the breathing of the patient is unaided, spontaneous or controlled by the ventilator, a mechanical breath. PEEP is used during mandatory mechanical ventilation when the breath is controlled by the ventilator and CPAP is the term used when the patient is still connected to the ventilator, but is breathing independently, spontaneously. CPAP is also often used in sleep therapy in order to keep the airways open to treat sleep-related breathing disorders including sleep apnea.

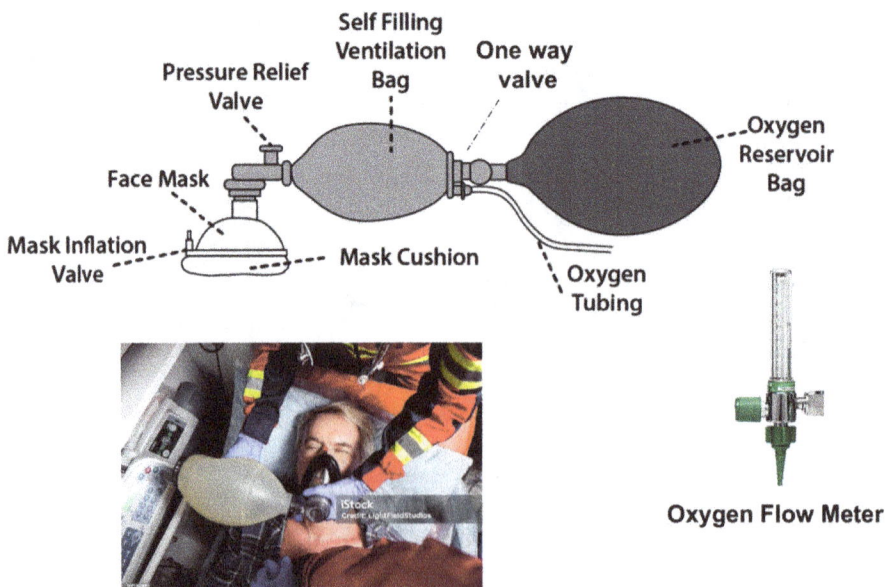

FIGURE 4.2.5   Bag, valve and mask (AMBU bag).

The operation of an AMBU bag is a simple but very effective means of quickly ventilating a patient. Figure 4.2.5 shows the components that make up the AMBU bag circuit. Starting at the right-hand side of the diagram there is often, but not always, an oxygen reservoir bag that is attached if supplemental oxygen is required, and this is fed with 100% oxygen via a flow meter. Oxygen supply to the flow meter can be either from a gas wall outlet or a gas cylinder. Supplemental oxygen is very often required due to the patient's low oxygen arterial blood saturation which is measured by a pulse oximeter.

The self-filling ventilation bag (SFVB) is constructed in such a way that it always returns to a predefined shape of a rugby ball following it being squeezed by the clinician. As the self-filling bag returns to its predefined shape, it will draw either oxygen from the oxygen reservoir bag or room air if the oxygen reservoir bag is not fitted. The face mask must be secure and snugly fitted to ensure that there are no leaks, and it can be secured by the fitting of a head-strap around the patient's head. It is also possible to connect the AMBU bag circuit via a connection to an endotracheal tube if the patient already has one fitted. During the inspiration phase, as the SFVB is compressed, the mask inflation valve closes to atmosphere and allows the fresh gas to flow and the pressure in the patients' airways and lungs to reach a pressure limit set on the pressure relief valve. On completion of the inspiratory phase the clinician releases the squeeze on the SFVB, the pressure reduces, allowing the patient to expire airway gas back up through the now open mask inflation valve to the atmosphere. If needed, a PEEP valve can be fitted to the system in order to maintain that the lungs and airways remain open at the end of inspiration.

- Operating Theatre/Room
  - Manley, Blease and Nuffield
    - Time Cycled, Pressure Limited, Constant Flow
    - Standing Bellows, Hanging Bellows or Piston Driven
  - Integrate within Anesthesia Machine
    - Bag in bottle eg. Datex-Ohmeda Aestiva/5 Anesthesia Machine
    - Full electronic Control ventilators with similar features to CCU ventilators

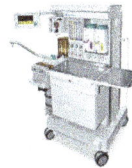

FIGURE 4.2.6   Operating room ventilators.

The very earliest use of mechanical ventilators occurred in the OR in order to sustain a patient's ventilation and they were simplistic mechanical ventilators that had just one mode of operation—controlled mechanical ventilation (CMV). This meant that they required the patient to be completely paralyzed and dependent on the ventilator for breathing. The breath was initiated by the machine, the inspiratory time period and expiratory time period controlled by the machine along with the maximum airway pressure. The speed of the constant air flow to the patient was also controlled by the machine. This method of ventilation often meant that a one-size fits all approach was often adopted, leading to issues such as barotrauma (over-pressure within the lung causing damage to airways and alveoli). It also impacted on the cardiovascular system, reducing cardiac output. The lack of ability to monitor such parameters as the patient's carbon dioxide (capnography) and arterial oxygen (saturated pulse oximetry, SpO_2) meant that the anesthetist was unaware of the negative impact on the patient's condition that the ventilator was having.

Today the ventilators found in the ORs are fully incorporated within the anesthesia machines and are not stand-alone devices as they once were. The early machines were time-cycled, pressure-limited, constant flow machines that were either piston-driven or standing bellows or hanging bellows. The use of pistons to drive the air into the patient is still a technique used today in some modern anesthetic machine ventilators such as the Drager Primus, which is far more responsive to the patient's needs and demands. The use of corrugated rubber bellows (standing or hanging) to drive gas into the patient is well established and used by numerous manufactures (Figure 4.2.7). The principle of operation is that the corrugated bellows sit within a clear plastic/glass jar that is visible to the anesthetist and shows clearly the patients and ventilator are working together. A detailed description of principals of the modern OR ventilators can be found further in this book in the chapter on anesthesia.

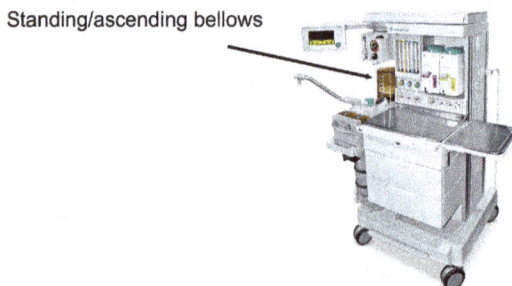

Standing/ascending bellows

FIGURE 4.2.7   Operating room ventilators – Bellows drive.

What follows is a brief description of how a standard OR ventilator delivers gas to the patient. Air, oxygen, anesthetic gasses and vapors within the corrugated bellows is driven into the patient when drive-gas (compressed air) is delivered within the surrounding space between the clear plastic/glass jar and the corrugated bellows. This results in the bellows being compressed, forcing air within the bellows to the patient. The drive-gas pressure within the bottle is then held for the duration of the inspiratory phase and is then vented to the atmosphere, allowing the patient to start the expiratory phase that returns the bellows to the top of the jar.

Hanging/Descending or Standing/Ascending Bellows?

Hanging/descending bellows were at one time regularly adopted by manufactures as they were seen as easier for the patients to expire against. However, there were some issues regarding to the detection of leaks and accurate delivery of set tidal volumes. Today many manufactures have more often adopted standing/ascending bellows as these have certain advantages. The first is that at the start of the inspiratory phase the bellows can be seen by the anesthetist to have reached the top of the clear bottle/jar. Markings on the side of the bottle show the volume delivered as the bellows are compressed by the drive-gas. During the expiratory phase the patient breathes the air back into the bellows and returns the bellows to the top of the bottle. If there were to be a leak to atmosphere in any of the connections between the patient's airway and the corrugated bellows, then the bellows would not fully return to the top of the bottle/jar. One other advantage is that the weight of the bellows means that there is always a minimal amount of PEEP created in the ventilator circuit that provides support for keeping the small airways and alveoli open.

- CCU/ITU/SCBU
  - Siemens, Drager and Puritan Bennett and others
  - Sophisticated Modes
  - Therapy in mind
  - Long term ventilation
  - Modern units have no bellows

FIGURE 4.2.8   Critical care ventilators.

Today's modern CCU ventilators are among the most advanced devices that you will find in any hospital (Figure 4.2.8). Major companies such as Siemens, Drager, Puritan Bennett, GE, Hamilton, Resmed, Medtronic and Mindray offer ventilators to the CCU hospitals, in a market that is worth more than $4 billion annually. Each company devotes a vast amount of money to continually improving current modes of ventilation and developing new modes. Primarily, CCU ventilators do not just provide ventilation, but just as importantly, they are a therapy tool in order to improve the patient's overall condition. Crucially, successful ventilation needs to provide sufficient air volume with the minimum of airway pressure. Our normal mode of ventilation, when breathing spontaneously, is negative pressure ventilation, as the diaphragm and intercostal muscles create a negative pressure around the lungs. It is important that positive pressure is always kept to the minimum required, but this is not always possible as some patients will need additional pressure to overcome air flow resistance due to such conditions as COPD and ARDS (acute respiratory distress syndrome), within the alveoli fluids begin to accumulate, and other respiratory diseases and injuries.

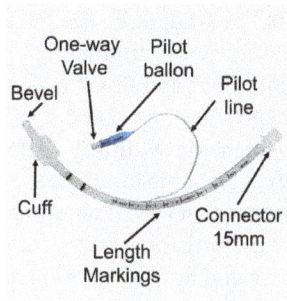

The cuff is used to ensure a seal between the trachea and the ET tube. Thus all gas must travel within the ET tube and none are allowed to escape to the upper airways.

This is vital for correct operation in "Volume Ventilation"

FIGURE 4.2.9

In discussing CCU ventilation, it is important to note that there are standard ways in which the air to the patient can be controlled. Given that a patient has had an endotracheal tube fitted in order to deliver air to the lungs, one of two methods can be selected. The first is pressure ventilation, where the air is delivered via the ET tube to the lungs to a predetermined pressure set on the ventilator. As the air leaves the end of the ET tube, the lungs inflate and any additional air pressure can escape along the gap between the ET tube and the trachea. This then escapes to atmosphere via the mouth and nasal passages.

The second method is volume ventilation. This method involves securing a seal between the ET tube and the trachea so that air only flows in and out of the patient via the ET tube (Figure 4.2.9). This method is often selected for the more sedated patients. To provide the seal between the ET tube and the trachea, a small cuff is inflated, close to the lower end of the ET tube. The procedure for inflating the cuff is as follows: a small syringe filled with air is connected to the one-way valve, the syringe injects air through the one-way valve that then delivers the air to the cuff via the pilot line. The cuff then inflates between the ET tube and the trachea. Once a full seal is achieved, the additional pressure in the pilot line causes the pilot balloon to inflate, which indicates that the cuff is fully inflated.

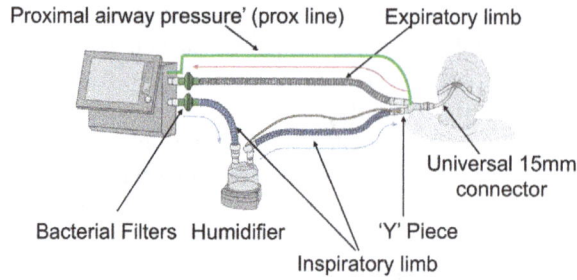

FIGURE 4.2.10 Typical ventilator circuit.

From the ET tube there is often a short piece of universal 15 mm tubing/connector that then connects to the 'Y' piece (Figure 4.2.10). On some ventilators you may also see a tube that is connected to the top of the 'Y' piece via a small connection port and narrow piece of tubing that connects to a port on the front of the ventilator. This line is used to measure the proximal airway pressure (prox line) that indicates the airway pressure within the ET tube. One thing you may notice is that the expiratory limb tubing and inspiratory limb tubing are corrugated tubings which are light, almost clear and flexible to maneuver. Being corrugated means that it does not expand during the inspiratory phase and thus delivers the pressure and air volume to the patient without loss due to expansion within the tubing. Two bacterial filters are usually connected in order to protect against the transmission of harmful infections between the machine to patient and patient to the machine and local environment.

Internally within the ventilator are both the expiratory and inspiratory valves along with the airflow generator, pressure and flow monitoring devices. Manufactures have, over the decades, developed a number of ways to produce airflow generation. These range from turbine, high precision controllable fans and banks of solenoids in which each solenoid when opened contributes a given amount of airflow to the total required flow. There are also ventilator that use a piston to also generate high airflows.

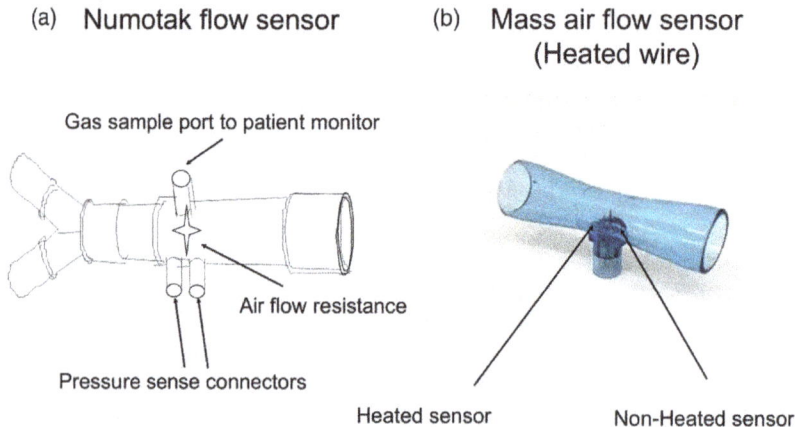

FIGURE 4.2.11 Airflow sensors.

These sensors are an important component in controlling the delivery of flow, volume and pressure of air to the patient. The measurement devices often used to measure the air flow is either numotak or mass air flow sensors (Figure 4.2.11). The numotak sensor is constructed using a tube with a known airflow resistance placed in the center of the tube. As the air passes down the tube and encounters the resistance, the pressures on the two sides of the resistance will differ. The numotak flow sensor has two ports, one on either side of the internal resistance. These two ports are connected back to the ventilator/patient monitor and the pressure on each side is reflected back up the narrow connecting tubes, back to two pressure transducers within the ventilator or monitor. As a result of knowing the value of the resistance in the center of the adaptor and the pressures on either side of the resistance, the ventilator software is capable of calculating the flow rate and therefore, by multiplying it by time, the volume.

A mass air flow sensor (heated wire) works by having two sensing wires in the center of the tube. One is heated and the second sensor is not. When air flows across the sensors they are both cooled. Electrical current is then increased or reduced to the heater sensing wire in order to match the unheated sensor wire. The amount of current increase or decrease is monitored by the ventilator software and airflow rates and volumes calculated.

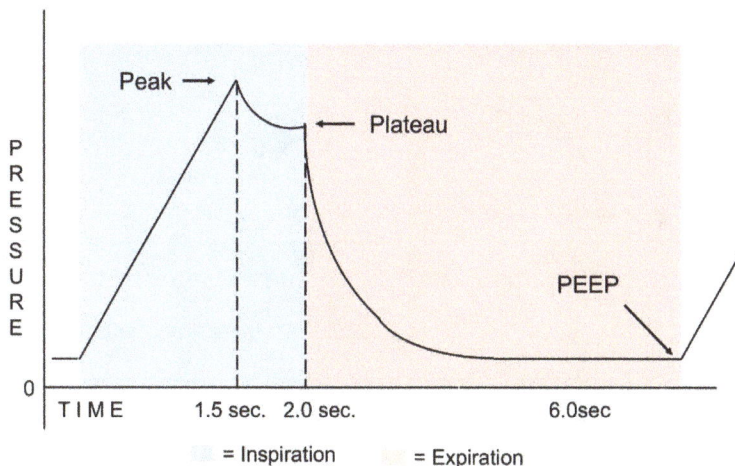

FIGURE 4.2.12 Ventilator pressure waveforms.

We now turn our attention to basic pressure ventilation that might be seen when ventilating a completely respiratory-paralyzed patient. The first thing to note is the total length of the breath shown is 6 seconds, meaning that the respiratory rate has been set to 10 bpm. This has then been divided into 2 seconds for the inspiratory period and 4 seconds for the expiratory period, so there is a ratio of 1:2 of inspired to expired.

This is about the ratio we all breathe during normal spontaneous breathing. It would be understandable if you had thought that our usual breathing pattern was something akin to a sinusoidal waveform, where we breathe in and out in equal time. As a small experiment, I would ask you to try for yourself breathing in and out in an equal time sinusoidal pattern. It should soon become apparent that this is very difficult and certainly not a natural way to breath. You will find that regardless of the breathing rate set on a ventilator, the default ratio is almost always 1:2. For certain medical conditions, the ratio can be changed and even inverted to give a longer inspiratory time than expiratory time.

Figure 4.2.12 is for a patient whose ET tube has had its inflatable cuff inflated, meaning that air can only pass in and out through the center of the ET tube and not escape around the sides of the ET tube. During the inspiratory phase the pressure has built to a maximum taking 1.5 seconds. What has then happened, are both inspiratory valve and expiratory valve has now been closed, trapping the air within the lungs. Due to the pressure held within the lungs, the air in the lungs begins to dissipate further throughout the lungs into the furthest reaches of the alveoli sacs, which leads to a drop in pressure from the peak pressure to a plateau pressure. This drop in pressure is to be welcomed because a bigger drop in pressure indicates a more compliant (elasticity of the lungs and chest wall) set of lungs. This is an excellent indicator of a healthy set of lungs. Diseased lungs will usually have poor compliance and thus show only a minimal drop in pressure between peak pressure and plateau pressure.

The expiratory phase starts at the end of the inspiratory phase with the opening of both the expiratory valve and inspiratory valves. The expiratory valve is only partially opened and will still offer some small resistance to the flow of air from the patient and the fresh air being supplied continuously from the ventilator. You might think of this as the feeling you would get if you placed your head out of a car window facing forward as the car is driven along. There is a small amount of opposing pressure to your attempt to breath out. At the end of the predefined time for the expiratory phase the pressure reading for PEEP is taken. An expected pressure reading of 1–7 cmH$_2$O is usual. For some respiratory conditions no or little PEEP is required. What then follows the end of the expiratory phase is the inspiratory phase for the next breath.

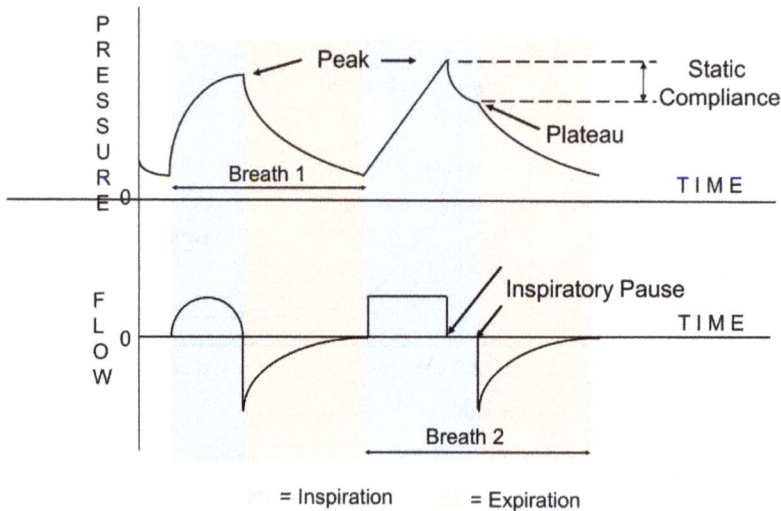

FIGURE 4.2.13 Ventilator waveforms flow patterns and static compliance.

Continuing on from Figure 4.2.13, we can see how the flow delivery of air can produce significant changes in the pressure waveforms.

For **breath 1**, we see that the flow delivery is rising in a smooth rising pattern and then begins to decelerate back toward zero flow as the pressure reaches its peak. This pattern of flow delivery is similar to how we would breathe spontaneously, and there is no inspiratory pause. At the point that the peak pressure is achieved, the ventilator then opens the expiratory valve and the patient then expires the air. As has been stated before, the expiratory phase is controlled by the patient's diaphragm and intercostal muscles working together to expel the air. The pressure is not allowed to fall to zero as a setting of a few cmH$_2$O PEEP has been set on the ventilator in order to ensure that the small airways and alveoli remain open and slightly inflated and do not collapse.

Breath 2 has a very different air flow pattern; at the start of the inspiration phase, the ventilator immediately delivers a preset flow rate of air which produces a rapid rise in pressure to preset pressure limit. Once this pressure limit has been archived, the air flow immediately ceases, and the inspiratory valve closes and the expiration valve remains closed. This means that at this point the inspired air is held within the lungs and begins to distribute itself further into the lower lobes. As a consequence, the peak pressure then falls to a lower level. This drop is indicative of the static compliance (measured when there is no air flow) of the lungs. Static compliance is measured, and it is thought that the greater the drop, the healthier the lungs are. The formula used to calculate the Static compliance is the tidal volume/plateau pressure—PEEP—and is expressed in ml/cmH$_2$O. By extending the period of time of the inspiratory pause, more time is given for the air to distribute further within the lungs until a stable reading of plateau pressure is achieved.

$$\text{Dynamic Compliance} = \frac{\triangle \text{Volume}}{\triangle \text{Pressure}}$$

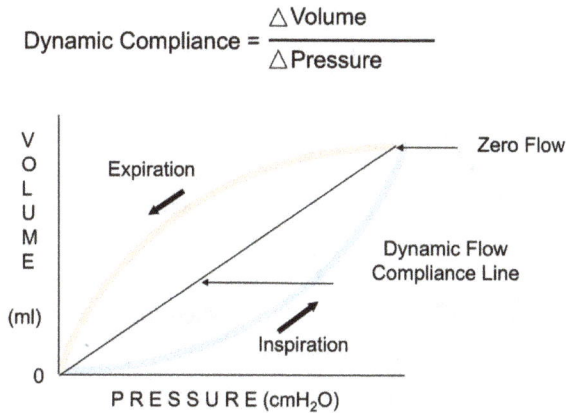

FIGURE 4.2.14   Pressure volume loop – dynamic compliance ventilator spirometry.

VENTILATOR SPIROMETER

The use of flow, volume and pressure loops are a visual tool that enables the clinician to gain a deeper understanding of the lung's response to positive pressure ventilation (Figure 4.2.14). As the flow of air enters and leaves the lungs during respiration, the clinician is able to see the effects as air is distributed deeper into the lungs on inspiration and also ascertain during expiration where resistance to the expiratory flow might be occurring.

Dynamic compliance (measurement calculated as the air continuously flows). This is calculated by drawing a line between the two points of zero flow, that is, the point before inspiratory flow begins and the transition from inspiration to expiration. The points along this line represent the changes in volume over the changes in pressure. The angle of this line is a visible indication of the dynamic compliance of the lungs.

On many modern ventilators and patient monitoring systems, it is possible for the clinician to take a snapshot of a loop at a particular time and then recall it to the screen later to compare the change in compliance value and shape of the loop against the current loop. This feature allows the clinician to constantly evaluate changes in the patient's lungs compliance and respond, if necessary, with changes to ventilator settings or therapy.

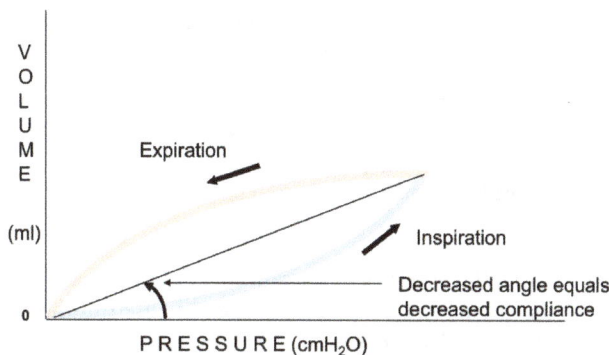

FIGURE 4.2.15   Pressure volume loop—decreased dynamic compliance.

As the position of the loop has moved more to the right, this indicates that in order to deliver a lower volume, then increased peak pressure is required and so the patient's lung compliance has decreased (Figure 4.2.15). This situation can be caused by many factors such disease, a fluid build-up within the lungs and adverse reaction to some anesthetic agents that cause a rise in surface tension within the alveoli.

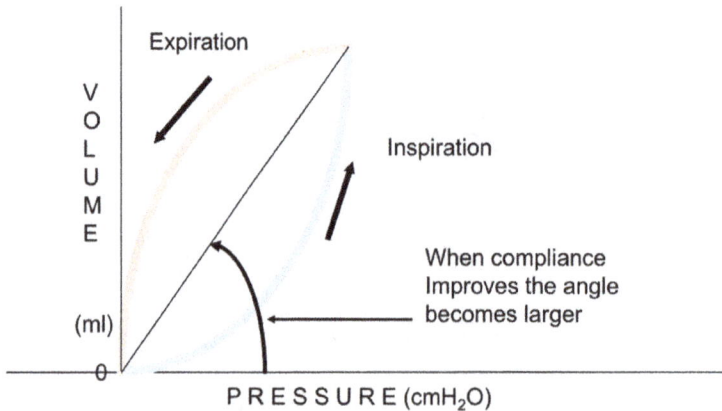

FIGURE 4.2.16   Pressure volume loop – increased dynamic compliance.

With improving compliance the angle will become larger, indicating a larger volume is being delivered for a reduced peak pressure (Figure 4.2.16). This of course is welcomed by the clinician as a positive indication of the improving condition of the patient's lungs.

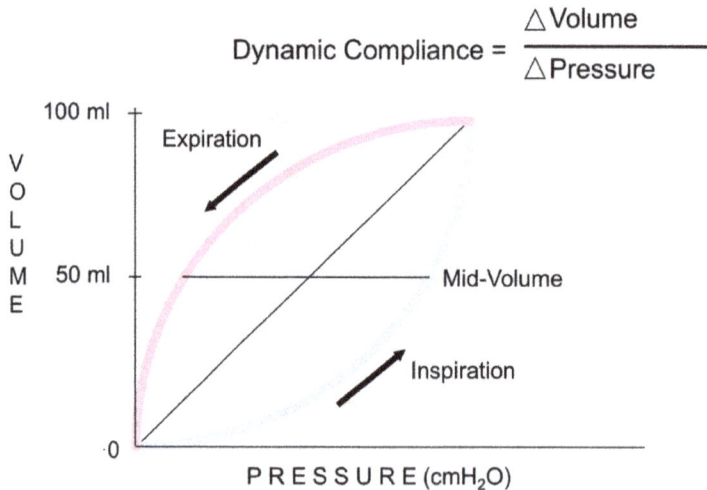

FIGURE 4.2.17   Pressure volume loop – mid-volume compliance measurement.

As can be seen in Figure 4.2.17, the dynamic compliance measurement does not require an inspiratory hold. Therefore, the mid-point of the delivered volume is used in the calculation along with the pressure measurement at that point to calculate the compliance figure. This is referred to as the mid-volume method.

Common Ventilator Terms

- WOB - Work of Breathing
- BPM - Breaths Per Minute
- Vt -Tidal Volume
- MV - Minute Volume
- I:E Ratio - Inspiration time to Expiration time Ratio
- PEEP – Positive End Expiratory Pressure
- CPAP – Continuous Positive Airway Pressure
- PSV - Positive Pressure Support Ventilation
- TRIGGERING – Pressure and Flow

FIGURE 4.2.18

Summarized in Figure 4.2.18 and what follows in detail is a list of common ventilator terms.

WOB—Work of Breathing: This is a term that is used to express the ease or difficulty the patient has when breathing spontaneously. Such factors as elasticity of the lungs (compliance), airways restrictions and the body's metabolic need for oxygen are assessed and using an advanced mathematical formula a value is obtained that can be used in determining the improvement or deterioration in the patient's condition.

BPM—Breaths per Minute: The number of breaths in one minute that the patient takes either spontaneously, assisted or controlled by the ventilator.

Vt—Tidal volume: This is the volume of air that the patient inspires on a single breath, either a spontaneous breath or a mechanical breath. A volume of around 500–700 mL is normal for an adult.

MV—Minute volume: By multiplying the tidal volume by the breath rate, we arrive at the minute volume. This is often a combination of all three types of breaths—spontaneous, assisted and mechanical. In some modes the target minute volume is set, and the ventilator will provide additional breaths in order to achieve the set minute volume.

I:E Ratio—Inspiration Time to Expiration Time Ratio: This is the ratio that compares the inspiratory phase time to the expiratory phase time. A typical ratio of 1:2, inspired to expired, is commonly seen, but there are circumstances where even inverse ratios such as 2:1 can be set on a ventilator in order to assist certain patients who may have such conditions as acute respiratory distress syndrome (ARDS – a type of respiratory disorder).

PEEP—Positive End Expiratory Pressure: This is a term that only relates to controlled or assisted modes by the ventilator. It is a baseline pressure that is maintained and measured at the end of the inspiratory phase. Its main purpose is to ensure that the alveoli and small airways do not collapse during the expiratory phase.

CPAP—Continuous Positive Airway Pressure: This is a baseline pressure that is continuously supplied by a ventilator or CPAP machine to a patient who is spontaneously breathing for themselves without ventilator support. Again, it is provided to ensure that the alveoli and small airways do not collapse during the expiratory phase and throughout the breath cycle. It also assists the patient during the inspiratory phase. CPAP is also mode of ventilation that is used for both intubated and none-intubated (i.e., facemask) patients who are breathing for themselves spontaneously. It is also seen outside the critical care areas of a hospital even in the home for patients with severe respiratory diseases such as COPD.

PSV—Positive Pressure Support Ventilation: This mode is only available on the patient's own spontaneous breaths. Its underlying purpose is to compensate for the restriction of the airflow in the inspiratory phase as the patient pulls air through the ET tube. As the ET tube is somewhat narrower than a trachea, it naturally causes a restriction that can be thought of as breathing through a straw. The effect of this is to increase the WOB for the patient as they draw air in during the inspiratory phase. What the ventilator will do to compensate for this restriction is that it provides a high addition flow/pressure during the initial start of the inspiratory phase that gradually reduces as the patient reaches the required tidal volume toward the end of the inspiratory phase. The benefit of this is that patients WOB is dramatically reduced, and the patient will feel as though they are breathing in as they normally would, with no restriction to airflow.

TRIGGERING—Pressure and Flow: The ventilator has two methods by which it is able to detect if the patient is making an effort to take a spontaneous breath. The first is seeing a sudden drop in pressure from the base line pressure, often from the PEEP base line pressure; and the second is seeing a small but rapid increase in flow as the patient pulls additional air for spontaneous breathing. Trigger settings on the ventilator allow the clinician to set a value for either pressure drop, and flow increase on the ventilator before the ventilator will deliver the additional mechanical breath. This setting of trigger values means that as the patient improves, the trigger value can be raised in order to encourage the patient to make more of an effort in order to receive an additional mechanical breath. In gradually increasing the effort the patient must make in order to receive a mechanical breath, this is used to start the process of 'weaning' the patient off the mechanical ventilator to start spontaneously breathing for themselves.

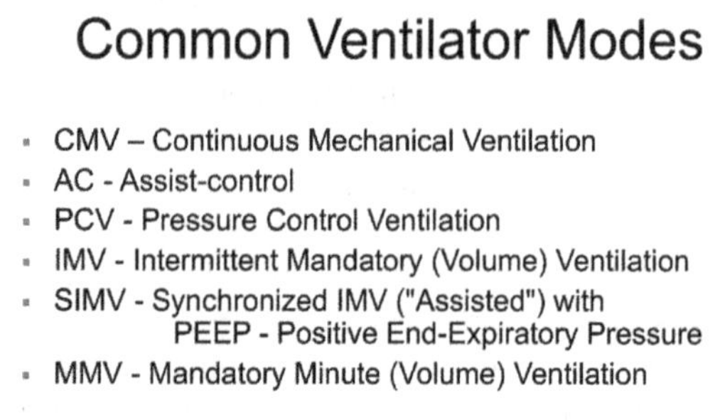

Common Ventilator Modes

- CMV – Continuous Mechanical Ventilation
- AC - Assist-control
- PCV - Pressure Control Ventilation
- IMV - Intermittent Mandatory (Volume) Ventilation
- SIMV - Synchronized IMV ("Assisted") with
            PEEP - Positive End-Expiratory Pressure
- MMV - Mandatory Minute (Volume) Ventilation

FIGURE 4.2.19

Ventilator mode refers to the way a mechanical ventilator will operate in order to support a patient's breathing. Summarized in Figure 4.2.19 and what follows in detail is a list of ventilator modes that can be applied to a patient who at first is in complete respiratory collapse, all the way through to full spontaneous normal breathing. This is not to say that throughout their recovery all of these various modes will be deployed. Also, I should point out that this is by no means a complete list of all the modes that are available on all patient ventilators, it is just some of the most common modes available.

The patient may be in full respiratory collapse for the following three modes:

CMV—Continuous Mechanical Ventilation: This is a mode of ventilation where all of the patient's breaths are initiated, controlled and terminated by the ventilator. This is a mode of 'full' support to a patient that makes no respiratory effort whatsoever.

AC—Assist-Control Mode: Is a mode in which the breaths are initiated by both the patient and the ventilator. This is a volume mode of ventilation, meaning that the cuff on the ET tube is inflated. When the patient attempts to inspire, this is then seen by the pressure sensors within the ventilator as a drop in the baseline pressure and the ventilator then responds by delivering a breath to a set tidal volume (Vt). If the patient does not make any respiratory effort, then the ventilator will initiate, control and terminate the required number of tidal volume breaths set on the ventilator in order to achieve the set minimum minute volume.

PCV—Pressure Control Ventilation: PCV is a commonly used mode when the patient does not have the ET tube cuff inflated (uncuffed). There is a preset target pressure, PIP, that is set on the ventilator. The air flow is generally high at the start of the inspiration phase and then slows as it reaches the target pressure. Excess air will leak past the outer side of the ET tube and escape via the mouth and nose. One of the advantages of PCV is that if the patient coughs or there is airway obstruction, then any excess pressure is immediately vented and the patient is unlikely to suffer barotrauma (injury to the lungs due to excessive pressure).

For the use of the following modes, the patient may now be showing signs of recovery from full respiratory collapse to beginning to attempt to take spontaneous breaths for themselves.

IMV—Intermittent Mandatory Ventilation: In this volume mode, the patient should make a spontaneous effort to take a breath, even during the expiratory phase of a machine breath, the ventilator will recognize this and provide air flow to support the patient's spontaneous breath. This can be triggered by either a sudden small drop in pressure, pressure triggered, or a change in air flow speed, flow triggered.

SIMV—Synchronized IMV: This mode is similar to IMV but is also capable of adjusting the delivery time of the next mechanical breath following a patients spontaneous breath in order to achieve the required minimum number of breaths and the set tidal volume. What is often seen is that the patient's spontaneous breaths are smaller in duration and volume than the mechanical breaths.

MMV—Mandatory Minute (Volume) Ventilation: It is a combination of SIMV and PSV in order to maintain a set MV by constantly adjusting the synchronization of the machine delivered breaths between the patient's own spontaneous breaths.

Other Common Ventilator Modes and Terms

- PRVC - Pressure Regulated Volume Controlled (volume preset pressure ventilation; machine alters pressure on a breath by breath basis to generate the tidal volume set by the user)
- BiPAP/BiLevel - Two-level CPAP

FIGURE 4.2.20

PRVC—Pressure-Regulated Volume Control: This could be thought as a mixed mode of ventilation by which the maximum pressure, PIP and inspiration phase time are adjusted breath-by-breath in order to ensure that the set required tidal volume (TV) is delivered. It has now become one of the most common modes of ventilation when a patient is admitted to the ICU department.

BiPAP/Bilevel—Two-Level CPAP: This is a noninvasive mode of ventilation, meaning that the patient is not intubated, and a face mask is usually used. During the expiratory phase, just as with CPAP, the lower additional pressure is provided to ensure that the alveoli and small airways do not collapse during the expiratory phase. From the start of the inspiratory phase, the ventilator provides an additional higher pressure in order for the patient to more easily breath in and thus reduce the inspiratory effort (WOB).

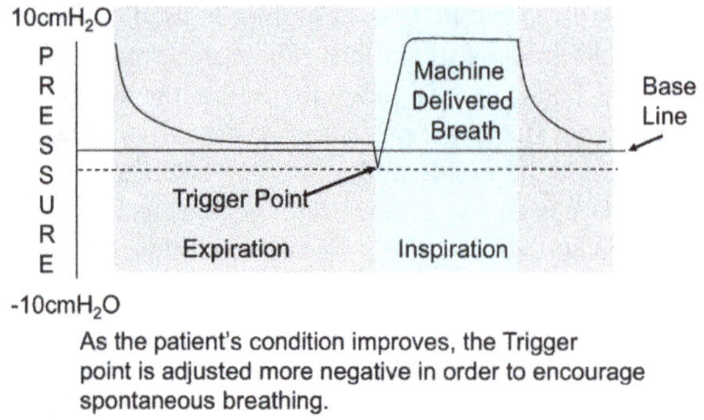

As the patient's condition improves, the Trigger point is adjusted more negative in order to encourage spontaneous breathing.

FIGURE 4.2.21   Pressure triggered.

Figure 4.2.21 shows that toward the end of the expiratory phase the patient attempted to inspire which resulted in the base line expiratory pressure falling rapidly to a negative pressure. This negative pressure trigger level may have been set on the ventilator as the point at which the ventilator will respond by delivering a machine breath. At first, during the patient's early recovery the negative-level pressure that the patient has to achieve is minimal, say -1 cmH$_2$O below the base line, but as they improve, their respiratory effort in order to trigger a machine breath is negatively increased, say to -5 cmH$_2$O.

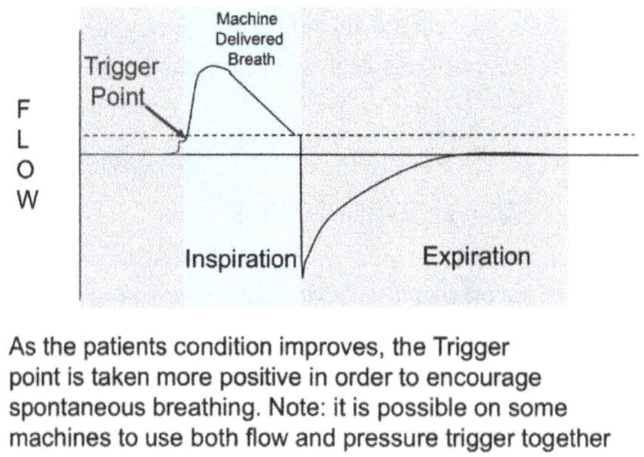

As the patients condition improves, the Trigger point is taken more positive in order to encourage spontaneous breathing. Note: it is possible on some machines to use both flow and pressure trigger together

FIGURE 4.2.22   Flow triggered.

An alternative to pressure triggered is flow trigger (Figure 4.2.22). The advantage to flow trigger is that pressure triggering has to develop the drop in pressure over resistance in order to activate the machine-delivered breath. As many patients may not have any or minimal restriction within their airway, it therefore can be difficult to produce a significant pressure drop in order to activate a machine-delivered breath. Flow trigger is activated by seeing a certain amount of extra flow, pulled by the patient, which once it breaches the flow trigger value set on the ventilator will then trigger the machine to deliver a breath. This method of triggering has the advantage that it is often much easier for a patient to trigger a machine-delivered breath.

Patient Weight (Kg)	BPM	Pinsp. Above PEEP (cmH$_2$O)	Inspiratory Time (sec)	Volume Limit (ml)
10	30	20	0.6	200
15	25	20	0.8	300
20	20	20	1.0	400
30	18	20	1.1	600
40	16	20	1.2	800
50	14	20	1.4	1000
60	12	20	1.6	1100
70	12	20	1.6	1200

FIGURE 4.2.23 Typical settings for pressure ventilation.

Figure 4.2.23 is a general guide to settings that maybe set on a patient ventilator that is being used in pressure ventilation mode. As you can see the breath rate reduces with the increase in the patient's weight, you will also observe that the inspiratory time and delivered volume increase with the increase in patient's weight.

Patient Weight (Kg)	BPM	Tidal Volume (ml)	Max Pressure cmH$_2$O
10	30	100	23
15	25	150	25
20	20	200	30
30	18	300	33
40	16	400	35
50	14	500	40
60	12	600	40
70	12	700	40

Question 1. For a patient of 50Kg, what would be the Minute Volume give the values above?

FIGURE 4.2.24 Typical settings for volume ventilation.

Figure 4.2.24 shows examples of settings that could be set on a ventilator when using volume mode. If it is at all possible, I would encourage, with permission, that you visit an ICU department and observe ventilators in use. Of course, you must always respect the patient's right to privacy and only concern yourself with the operation of the ventilators.

5 Anesthesia Principals and Terminology

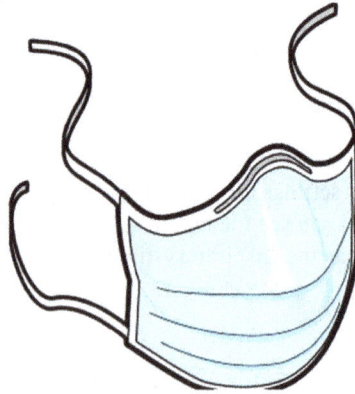

Anesthesia is generally considered very safe, but it is after all the administration of compounds and gases that in themselves could be considered poisonous. Unlike almost anything else in medical science, anesthesia is not directly therapeutic. In general, it is for the sole purpose of incapacitating a patient so that they do not respond to surgical procedures. Sometimes anesthetic agents are given for sedation only, in these cases the patient may be given time while unconscious to recover from disease or injury.

Many years ago, while I was undertaking training on the subject of anesthesia, a senior and highly experienced consultant anesthesiologist made the comment, 'You have one foot in heaven when you are under a general anesthetic'. This might sound flippant, referring to the very deep sleep that those under anesthesia experience, but it is also a reminder of the dangers that come with putting a patient under an anesthetic and the ever-present risk of death if any of those involved should make a mistake, or the equipment should malfunction.

Unlike many other procedures undertaken in the hospital environment, anesthesia is somewhat different in that it is not a sequential process. When a patient is in the operation room (OR) there are a number of factors in play at any one time, and each can have an impact on the others. The patient, the surgeon, the anesthetist, the patient monitor, the anesthetic machine, infusion devices, and the ventilator, all simultaneously work together to maintain the patient. A change in one can precipitate a change in the other factors.

As your experience increases, you may well find yourself working in the OR. This environment at first can be very intimidating as it is the preserve of highly trained clinical staff. A rule of thumb for working in the OR is that you should make clear who you are and what your role is, then only speak if spoken to. As your experience grows, you should look carefully at all that is happening and not just at the device you may have been called in to deal with. On occasions, clinical engineers attend OR after the surgery has been completed, only to find themselves testing medical devices and noting that they are unable to replicate problems that arose during surgery. It is best practice to

DOI: 10.1201/9781003609414-5

always attend immediately to problems in the OR in order that you can see for yourself the issue and speak with those who have reported the problems.

Anesthesia

=

Insensitivity to pain

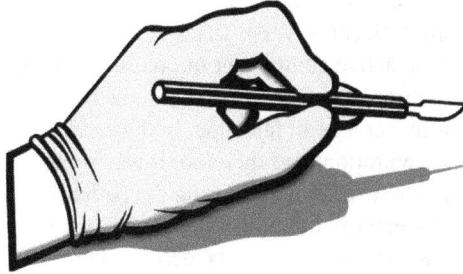

FIGURE 5.1.1

Anesthesia is insensitivity to pain (Figure 5.1.1). The Webster's dictionary definition is *'Loss of sensation with or without loss of consciousness'*.

Anesthesia is therefore a wide and varied subject. It encapsulates everything from an aspirin tablet all the way through to the most sophisticated anesthetic practices found in the OR. Primitive anesthesia practiced for thousands of years includes such practices as the use of acupuncture, opium and alcohol in attaining pain relief.

Anesthesia, a Modern Historical Review

Discovery of oxygen	1770s
Nitrous oxide as "laughing gas"	1808
Discovery of morphine	early 1800s
Use of ether & chloroform	early 1800s
First general anesthesia (with ether)	1846
Main blood groups: transfusion of blood possible	early 1900s
First intravenous anesthesia – (Pearl Harbour)	1930's & 40's
Muscle relaxants	1940
Manually controlled breathing	1940s
Modern inhaled anesthetics	1956 -

FIGURE 5.1.2

Figure 5.1.2 lays out some of the most significant discoveries in the modern history of anesthesia to the present day. It is by no means a total comprehensive guide to all the discoveries that have contributed to the subject of anesthesia. In little more than 250 years, our understanding of chemistry, gases and compounds has come an enormous distance. It wasn't until the 1770s that oxygen was identified and the part it played in supporting combustion and assisting in breathing was recognized, first observed by Joseph Priestley. Even then, the significance to supporting life was not fully understood.

Possibly the first significant step in providing pain relief, as we know it today, was the discovery of nitrous oxide, sometimes known as 'laughing gas'. Joseph Priestley went on to demonstrate his discovery of nitrous oxide in 1772, but it wasn't until a dentist named Horace Wells demonstrated it as an anesthetic agent during a dental extraction in December 1844, and its medical properties were accepted and recognized as a significant anesthetic agent.

During the early years of the 19th century, Frederick Wilhelmina Sertürner, a German pharmacist, became the first to distil the active component in opium that he then called morphine. In the early years, this analgesic was administered orally, but with the invention of the hypodermic needle, it soon became possible to administer it as an injection. In the early years there were many problems with the consistency of the concentration, and therefore it was somewhat problematic in providing sufficient pain relief without severe over or underdosing the patient.

The advent of chloroform and ether in the mid-19th century was a major step in the development of modern anesthesia. These two substances could be delivered to the patient as vapors, often using what is called a Schimmelbusch mask. This mask was a simpler arrangement consisting of a piece of gauze, on top of a wire ring that was placed upon the patient's mouth and nose. Liquid ether or chloroform was then dripped upon the gauze and due to the patient's breathing they would inhale the vapor. This arrangement had many serious drawbacks, not least the fact that the clinician administering the ether or chloroform would also be very likely to inhale the vapor as well as the patient. Vital signs monitoring during such a procedure was a very simple assessment of the patient by visually monitoring the depth of respiration, the patient's pupils and any response to the surgical incisions.

One of the most significant step forwards in surgery and anesthesia occurred in 1900 when Karl Landsteiner identified the main blood groups of A, B and O blood types. From this time, it became possible to undertake blood transfusions safely by giving the patient the correct matching blood.

The 1930s and 1940s saw the advent of intravenous (IV) anesthesia, particularly in the United States. Based on administering opiate-based drugs, such as sodium panthenol by IV injection, it quickly became a refined method of anesthesia and was initially believed to be a safe method. However, mistakes made by inexperienced practitioners in administering barbiturates when treating the wounded during the attack on Pearl Harbor in December 1941 highlighted the potential risks, with anesthesia believing to have contributed to hundreds of deaths. During these years, the medical profession began to recognize that anesthesia was a separate and important clinical discipline. Up until this time most anesthesia procedures had often been performed by junior medical staff.

One of the three major elements of general anesthesia is that of muscle relaxant. It is absolutely necessary for successful, abdominal and thoracic surgery that the chest and abdominal muscles are flaccid in order for the surgeon to gain access to the inner organs. Curare, a drug obtained from the bark of certain trees, which had been used by the native peoples of many South American countries for the hunting of wild animals, proved to have the quality of inducing muscle relaxation, even to the point of paralyzing the heart and other muscles. The refining of curare into the more refined muscle relaxant drugs, we have today has taken many decades.

With the advent of the mechanical patient ventilator (respirator) around the 1940s, the depth of anesthesia could be increased to the point of respiratory paralysis, allowing for better pain relief and deeper levels of unconsciousness. It was for many decades in the OR that the ventilator was thought to be only for maintenance of the patient and not in itself therapeutic. Today many ventilators incorporated within the anesthetic machines have a highly sophisticated modes of ventilation that minimize the impact of anesthesia.

From the 1960s and 1970s, we have seen a rapid development in the anesthetic vapors that are used in the OR. Some of the early anesthetic agents used in vaporizers were often very unstable and even combustible and were known as volatiles. These were initially replaced with three well-known anesthetic agents—halothane, isoflurane and enflurane. In the last 20 years, these have also been replaced by the adoption of two more—sevoflurane and desflurane.

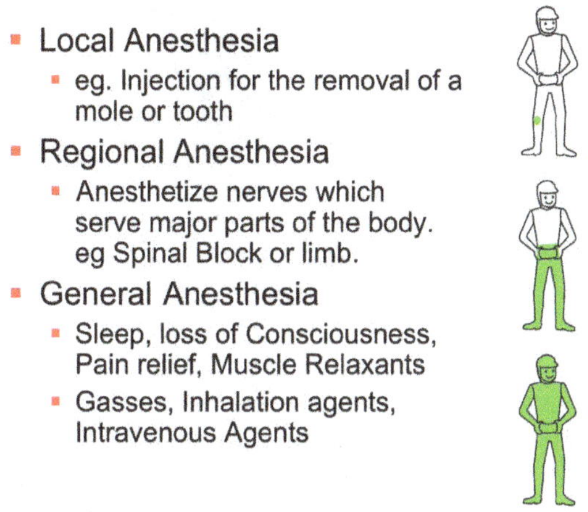

- **Local Anesthesia**
  - eg. Injection for the removal of a mole or tooth
- **Regional Anesthesia**
  - Anesthetize nerves which serve major parts of the body. eg Spinal Block or limb.
- **General Anesthesia**
  - Sleep, loss of Consciousness, Pain relief, Muscle Relaxants
  - Gasses, Inhalation agents, Intravenous Agents

FIGURE 5.1.3 Types of anesthesia.

The field of anesthesia can be divided into three areas (Figure 5.1.3). These areas match the surgical need against the need to minimize of the use anesthetic agents in order to provide sufficient pain relief, muscle relaxant and consciousness.

Local anesthesia: Possibly the most used anesthesia is that for a tooth extraction. Today the most common IV (intravenous) anesthetic agent used is Midazolam that is injected into the gum close to the sight of the tooth that is to be extracted. Not only does it give excellent pain relief, but it also helps to relax the patient and reduces anxiety. There are also a number of other commonly used IV anesthetic agents such as Epinephrine and Articane. Up until some 30–40 years ago, an anesthetic agent commonly used was nitrous oxide (laughing gas). This was a very effective anesthetic agent but had many drawbacks not least of all the pollution of the environment. For the removal of such things as a mole or wart, a commonly used anesthetic is Lidocaine that once used with other compounds can not only reduce pain, but also perfusion at the site where the mole is being removed. This helps in controlling the bleeding following the mole removal.

Regional anesthesia: This type of anesthesia is generally undertaken in the hospital environment and undertakes to provide pain relief over a larger area, such as a limb. One of the most used methods of this type of anesthesia is the spinal block where the anesthetic agent is inserted into the lower spine, providing pain relief from the waist down. Often during childbirth and particularly cesarean sections, (the baby is delivered through an incision in the mother's abdomen and uterus) or a delivery by forceps the mother will be given a spinal block. The procedure provides pain relief extremely quickly and has no impact on the wellbeing of the baby.

General anesthesia: This is almost always performed in the OR where there is availability of monitoring, ventilation and resuscitation equipment. Using a combination of anesthetic gases, volatile agents and IV anesthetic drugs, the patient is fully sedated and loses consciousness. In today's OR, it is possible to closely monitor the patient's pain, relief, consciousness and muscle relaxant by means of advance monitoring devices. A general anesthetic may last anywhere between 15 minutes and over 20 hours dependent upon the surgical procedures being undertaken. Several of the medical team working in the OR are involved with the safe delivery of anesthesia.

- A Balance of …
  - Sleep
    - Lack of Awareness
    - Amnesia
  - Relaxation
    - Lack of Muscle Tone
  - Analgesia
    - Pain Relief

FIGURE 5.1.4

Delivering anesthesia requires a balancing act that weighs the optimum levels of pain relief, muscle relaxant, consciousness and amnesia (lack of memory of the surgical procedure) against the need to only provide sufficient anesthetic agent to safely do so (Figure 5.1.4). Many factors work together in order to achieve this critical balance, such as the surgical procedure being undertaken, the patient's physical and mental condition, age and gender. You may think that it would be in the patient's best interest to be heavily anesthetized for almost all major surgical procedures, but this is not the case. What should always be kept in mind is that following the surgical procedures, the patient is recovering not only from the surgery but also from the effects of the anesthesia. As you progress through this chapter/presentation then you will discover how anesthesia has a detrimental effect on many of the body's organs and systems.

One simple example of how the level of anesthesia might be tailored is to try this small personal experiment. First, make a small fist with your hand. Second, lightly tap the top of your skull with your fist and in small increments increase the force of your tapping of the skull until it becomes uncomfortable, and then stop. At this point imagine (under no circumstances attempt to strike your face) the same amount of forceful blow being applied to an area of your face. You should now begin to understand that the different areas of the body have very different levels of sensitivity to pain and thus the level of analgesia given would need to match the level of pain sensitivity that the patient will experience from the surgery. At the opposite end of the pain spectrum, I would ask you to think of the level of pain that would be felt for the surgical removal of a toe nail. I'm am sure that even the thought of such a procedure being performed without a high dose of analgesia would give most people nightmares, but this procedure can be undertaken without any muscle relaxant or sedation.

From the examples above that a general anesthesia is not a 'one size fits all', but is tailored for individual patients.

1. Analgesia = pain relief

2. Unconsciousness

3. Amnesia = lack of memory of surgical experience

4. Muscle relaxation = immobilisation of the surgical field

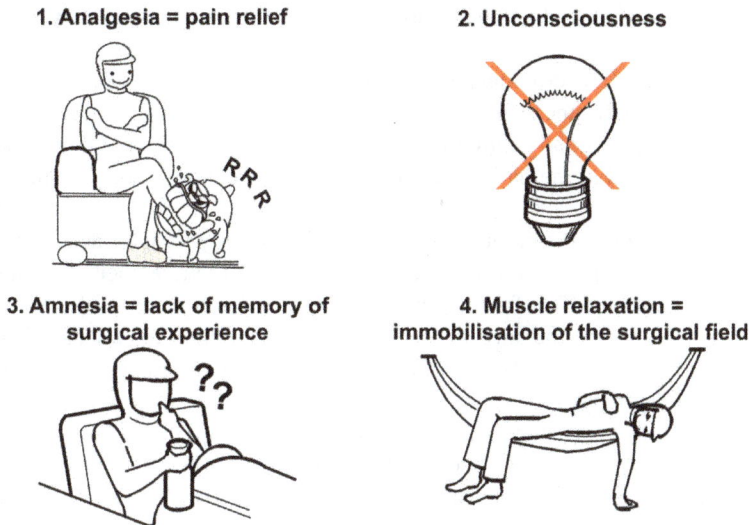

FIGURE 5.1.5   General anesthesia – properties.

Figure 5.1.5 should allow you to think of the need to consider how this balance is applied. Such as surgical procedures to the abdomen, chest, shoulder and hand usually involve muscle relaxants. Open heart surgery and intestinal surgery will most often involve deep sedation, unconsciousness and amnesia, as well as pain relief and muscle relaxants.

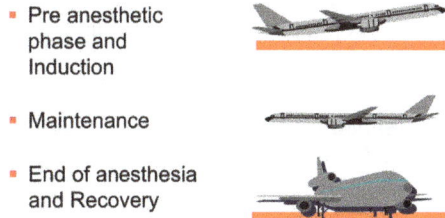

- Pre anesthetic phase and Induction

- Maintenance

- End of anesthesia and Recovery

FIGURE 5.1.6   General anesthesia – analogy.

One of the most useful analogies used to understand general anesthetic is that of taking a flight on a modern aircraft (Figure 5.1.6) Preparation for the flight might start at home with the packing of luggage and valuables you wish to take with you. In general anesthesia this usually means a pre-anesthetic assessment to check that you are physically able to not only with-stand surgery, but just as importantly, the anesthesia. Very often this can be undertaken a week before surgery or as close to surgery as the evening before when you've been admitted to a surgical ward. Of course, there are some occasions such as emergency surgery when this is undertaken just minutes before surgery.

The day of the flight arrives, and you are at the airport, proceeding through check-in and pass-port control. For general anesthesia, this might be the journey from the ward or reception area down to the operating rooms or even an anesthetic suite. Once aboard the flight you find your seat and fasten your seatbelt. In anesthesia, this might be where you are given a small injection in order to gently sedate you to the point where you are ready to be fully anesthetized. As the plane taxis down the runway, the pilots are busy with increasing the power to the engines in order to propel the plane at sufficient speed to take-off. This might be considered the induction phase where the patient is now, fully, anesthetized having been given a muscle relaxant, IV anesthetic and anesthetic gasses (volatiles). The process of an endotracheal tube being fitted (incubation) is then performed by the anesthetist.

The plane then climbs to its cruising altitude, and so in anesthesia the patient is monitored carefully as the paralysis from the anesthetic agents takes effect. Once reaching the cruising altitude, the plane will level off and pilots and cabin crew begin to relax at this point. The anesthetist will now have satisfied themselves that the patient has reached the correct depth of anesthesia and surgery may now commence.

The flight now proceeds to the point, where landing is now only minutes away. At this point, the pilot begins to reduce speed and everybody is asked to fasten their seatbelts, including the cabin crew, and prepare for landing. This for the pilots again is the period where high levels of concentration is required. At this point, the surgery has been completed and the patient is beginning to be withdrawn from the anesthetic agents. The anesthetist is carefully monitoring the patient's reaction and beginning to watch for signs of recovery such as the patient beginning to breathe spontaneously without the support of the ventilator.

As the plane lands the pilot and co-pilot are fully immersed in preparing the landing by reducing the plane's speed even more and engaging the wheels in the down position ready for landing. As landing happens, the pilots work to reduce the speed as much as possible while controlling the direction of the plane. With the patient recovering from the anesthetic, the patient awareness level begins to rise. Moment-by-moment monitoring of the vital signs such as blood pressure, temperature, SpO_2 and heart rate are undertaken to ensure that the patient does not suffer any relapse. At this point the patient is now becoming aware of his surroundings and anesthetic drugs are fully withdrawn. Extubating of the ET tube is now performed, and significant pain relief is administered if required.

From the flight the travelers disembark and make their way to the arrivals hall to collect their luggage and have their passports processed. For the patient they are moved from the OR and taken to the OR recovery area where they are again closely monitored for any adverse reactions before they are either discharged from the hospital or returned to a surgical ward.

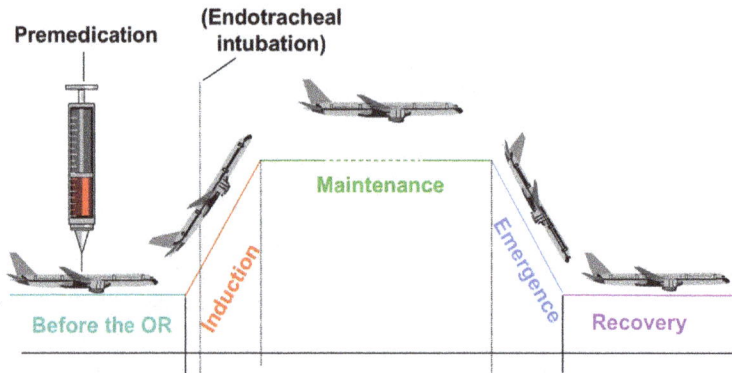

FIGURE 5.1.7 Anesthesia process.

Using the same flight analogy, we can look more closely at where issues in anesthesia might occur (Figure 5.1.7) As you may already be aware, it is very seldom that an aircraft suffers a catastrophic failure mid-flight. The occurrence of such catastrophic failures to aircraft's is less than 10% of all recorded catastrophic failures. So, it is take-off and landing that present the biggest challenge to safety. In general anesthesia this is very much the case that during induction and recovery the medical staff, particularly the anesthetist, monitor extremely carefully the reaction of the patient during these critical periods. There are occasions when a surgical procedure will be abandoned due to a patient's adverse reaction to the anesthetic induction process. Throughout the surgical procedure, the anesthetist continues to monitor the patient's vital signs, and if he observes a rapid change such as a decline in the patient's blood pressure, he will inform the surgeon immediately, and together they take steps to rectify the issue. This may even include abandoning any further surgery.

For clinical engineers working in the OR during surgery, there is one critical factor to keep in mind: it is expected that you will remain at some distance from the patient and be very quiet. Do not under any circumstances distract the anesthetist during these critical times.

FIGURE 5.1.8   Pre-anesthetic consultation

During the pre-anesthetic consultation, the anesthetist or a member of his anesthetic team will undertake an evaluation of the patient's general health in order to access any possible risks that might arise during surgery (Figure 5.1.8). The patient might be weighed, and blood pressure readings taken. It is usual that the risks are explained to the patient and any patient concerns are addressed. At the end of the consultation, the patient, parent or guardian are then asked to sign a declaration allowing for the surgical procedure to go ahead, and that they understood what the surgical procedure entails and the risks involved.

Patient Assessment

- Cardiovascular
  - Previous M.I. (Myocardial Infarction)
  - Hypertension (diastolic consistently >110mmHg
  - Valvular heart disease/prosthetic valves….
- Respirator system
  - Cough and production of sputum
  - Wheeze – COPD, Emphysema, Bronchiectasis, Asthma…..
- Family History
  - Hemophilia….

FIGURE 5.1.9

For major surgical procedures, during the pre-anesthetic consultation a detailed patient assessment is undertaken that involves not just a detailed review of the patient's own medical record, but also questions that relate to health issues of blood relatives such as parents and siblings (Figure 5.1.9). The anesthetist will usually also undertake to access the patient's current health and check for any of the conditions listed on the slide above. One question that is always asked is, has the patient had any adverse reaction to a general anesthetic before?

Patient Assessment Scale
The American Society of Anesthesiologists (ASA)

ASA 1	Healthy, normal, aged below 65
ASA 2	Mild systemic disease (e.g. mild hypertonia) or healthy, normal, aged over 65
ASA 3	Severe systemic disease limiting normal activity (e.g. diabetes with hypertension)
ASA 4	Incapacitating systemic disease which is a constant threat to life
ASA 5	Moribund, not expected to live 24 hrs without the operation
ASA (1-5 E)	Signifies that the operation is to be performed as an emergency. This places the patient in the next risk category

FIGURE 5.1.10   Patient assessment scale.

From the health data and information gained during the patient consultation and pre-anesthetic assessment, it is then possible to grade the patient against a scale such as the American Society of Anesthesiologists (ASA) patient assessment scale (Figure 5.1.10). This scale is commonly used in order to allow clinicians to grade the risks during the anesthetic and surgical procedures.

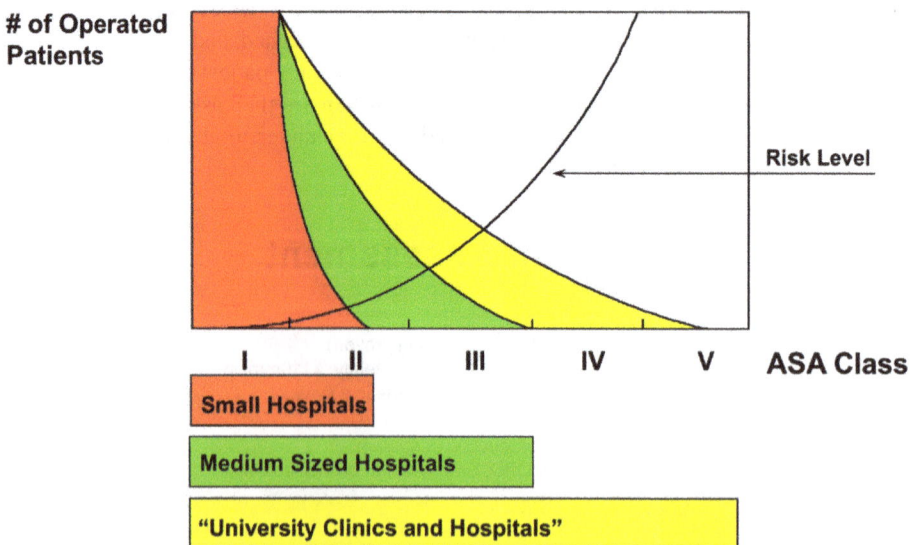

FIGURE 5.1.11   Patient assessment scale.

The purpose of Figure 5.1.11 is to convey how the patient's degree of risks is selectively managed when selecting where complex surgical and challenging anesthesia would be performed. For elective surgical procedures (nonemergency) that involve patients that have ASA classification scores of 4 or 5, it may well involve the patients being transferred to university hospitals where there may be a greater degree of expertise in handling potentially difficult anesthetic needs. Large university hospitals usually have not just more clinical experience, but also greater resources in terms of equipment and support services such as pathology and imaging equipment, that is, CT (computerized tomography) scanners and MRI (magnetic resonance imaging) scanners. In the event of a patient needing an emergency surgical procedure where time is of the essence, complex procedures may have to be undertaken at the smaller hospitals and the risks managed given the resources available.

- Gases
- Volatile/Vapours
- I.V. and I.M. Agents

FIGURE 5.1.12   Anesthesia dimension (delivery).

The methods used today of administering anesthesia are many and varied. Sometimes a single approach such as IV injections can be used, but often a mix of gases, volatiles/vapors, IV and IM (intramuscular) are carefully selected in order to attain the correct level of anesthesia for a particular patient and the surgical procedure being undertaken (Figure 5.1.12).

Yesterday's Anesthesia Machines

- 1925 The first Boyles Machine
  - A collection of Tubes, Regulators and Bottles
- 1950's The first Professional Boyles Machine
  - True integration of the various parts
  - Machines of this era often referred to as "Boyles Internationals"
  - The advent of VAPORISER

FIGURE 5.1.13   Evolution of anesthesia machines.

A look back at the history of today's anesthesia machines is worthwhile in order to gain an understanding of just how far these machines have evolved in the past 100 years (Figure 5.1.13). Many of the safety features that are standard on today's machines were not even considered back then. These early machines were not just hazardous to the patient but also exposed the medical staff exposing them to risks of contamination from the anesthetic agents that were being delivered to the patients. Certain anesthetic gasses and vapor combinations occasionally reacted with the rubber tubing and would produce dangerous noxious gasses that would be inadvertently delivered to the patient and the local atmosphere.

In order to administer the vapor, the gas delivery would pass through glass jars that had two tubes in the lid of the jar. In the bottom half of the jar was the liquid anesthetic agent such as ether. An inlet tube into the jar delivered the patient anesthetic gas, possibly oxygen and nitrous oxide to the top half of the jar. To vary the level of vapor concentration, the outlet tube within the jar would have had its height varied above the liquid. The closer the outlet tube was toward the liquid vapor, the higher the concentration of vapor in the patient anesthetic gas. The higher the outlet tube would be in the jar, the lower the concentration of vapor in the anesthetic gas would be. This of course would not be able to deliver accurate levels of vapor concentrations to the patient.

In the early 1950s, a serious attempt was undertaken to design anesthetic machines that were purpose-built with integrated systems for gas and vapor delivery. There was the adoption of gas flow tubes, which had flow rate markings on the side of the flow tube and with a bobbin within the tube that would rise as the flow rate through the tube increased. This could then be read against the flow rate marking on the tube in order to accurately measure the flow of what was being delivered to the patient. This period saw the advent of today's vaporizers, which were purpose built to deliver a calibrated dose of the vapor/volatile agents taking into account the gas temperature and atmospheric pressure.

Today's Anesthesia Machines

- Check list * Guideline from the A of A.
- Incorporated Safety features
- Oxygen failure warning
- Hypoxic guard
- Many others
- Minimum machine standards (British Standards)
- Minimum monitoring standards (A of A)

FIGURE 5.1.14

Each country will have its own set of standards covering the construction of anesthesia machines and the required minimum level of patient monitoring. Before any anesthesia and surgical procedure is undertaken, the anesthetist or staff will undertake to check the anesthesia machine, patient monitoring, emergency equipment, such as airway suction systems, and drugs. These checks are then recorded within the OR records system for possible review should there be an unexpected issue during the procedure(s). The importance of record keeping cannot be understated in the OR.

Today, checks are required on devices such as the oxygen supply failure alarm, which operates when the back mounted gas cylinder and/or the wall hose that is connected to the wall supply are disconnected. An independent gas-driven alarm whistle will sound for 10–15 seconds. Another such minimum machine standard is the hypoxic guard. This ensures that if nitrous oxide is being delivered to the patient, then a minimum of 25% oxygen is also delivered as nitrous oxide is not a substitute for oxygen. Other such safety features such as a pressure relief valve of the common gas outlet (CGO), checks of the flowmeters and vaporizers are also required (Figure 5.1.14). Checks of the scavenging system to ensure a safe OR environment are also undertaken.

With regard to monitoring, each country will also have a minimum standard before any general anesthetic procedure can be undertaken. These are commonly ECG, blood pressure, SpO_2, carbon dioxide and oxygen. They may also extend to respiration, temperature and in some advanced countries depth of anesthesia monitoring such as BIS (bispectral index). This uses EEG, the brains electrical activity to determine the depth of consciousness of the patient due to the anesthesia.

- Cape Nuffield and Manley
- Bag in bottle e.g. Standing/Hanging
- Electronic software control
- Pressure cycle
- Volume cycle
- ITU features ventilator

FIGURE 5.1.15 Ventilator types.

In order to obtain the most from the discussion on OR ventilators, it is strongly recommended that you first read the chapter on positive pressure mechanical ventilation.

From the 1950s onwards, there has been a steady development of mechanical positive pressure OR ventilators from Cape, Nuffield and Manley (Figure 5.1.15) to today's ventilators, which are integrated within the anesthesia machine. The original ventilators used in the OR were timed-cycled, pressure limited, constant flow machines, sometimes referred to as minute volume dividers, with only one mode of operation and limited safety features. The ventilator feature in the slide is a Blease Manley ventilator, which was typical of a 1960s/1970s machine. This for its time would be considered an advanced ventilator as it had an airway pressure monitoring module connected to the side. You will also observe that the delivered volume would have been set and measured by the distance the bellows moved against scaling on the curved arm that was attached to the top of the bellows.

The 1980s saw the increasing use of microprocessors and software in the construction of the OR ventilator. We also see at this time that OR ventilators are now becoming integrated within the anesthesia machines that provide for less complex external connections. Software control also allowed for the development of more advanced modes of ventilation such as pressure support ventilation (PSV) that greatly assisted the patient as they began to breath for themselves. Being able to select with ease, the various modes and settings within those modes enabled the anesthetist to easily tailor the ventilator settings to each patient needs throughout the surgical procedure.

Often, for longer more complex surgical procedures, volume cycle mode is selected, where the patient's ET tube is inflated ensuring that it is possible to accurately measure the volume of gas delivered to the patient in order to achieve a set 'minute volume' (tidal volume rate) with pressure limits.

For shorter surgical procedures, pressure mode is often used which allows for a higher flow rate but limited maximum pressure. In both modes there is usually a set value of positive end expiratory pressure (PEEP) set in order to ensure that the alveoli and small airways remain open.

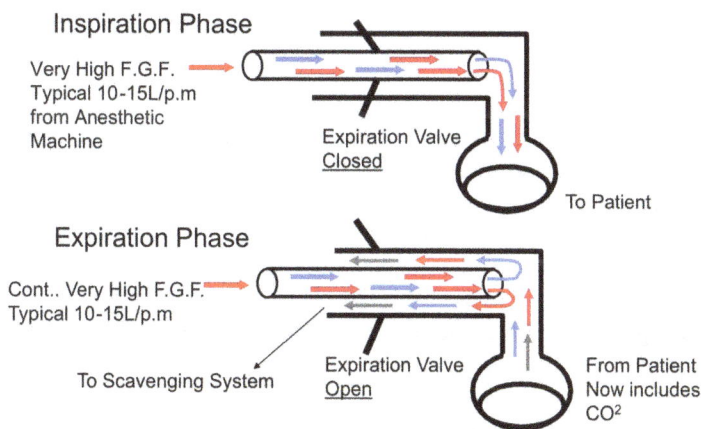

FIGURE 5.1.16   Induction breathing circuits (known as "TOP circuits").

Before the advent of the circle circuit system, a closed method of delivering the anesthetic gasses such as air, oxygen, nitrous oxide and vapors, the common methods of gas delivery were open circuit systems such as the open induction breathing circuit shown in Figure 5.1.16. This type of circuit has a number of drawbacks such as very high 'fresh gas flow' (FGF) rates.

- As you can see if the FGF rate is 10 L/min and the patient has an inspired to expired ratio of 1:2 then at 10 breaths/min (1 breath is 6 seconds long, of which inspired is 2 seconds and expired 4 seconds).
- The patient will only breathe in for a total of 20 seconds in any 1 minute. This means that for 40 seconds in any one minute the FGF will not enter the patient.

- • This of course means a loss of 66% of all the FGF in any one minute, as two-thirds of the FGF does not enter the patient.
- • Given that all of the FGF contains not just air and oxygen, but also nitrous oxide and vapor, then this represents a high degree of waste of expensive anesthetic agents and gasses.

However, one major advantage is at the induction stage, particularly with children, when it can be difficult to obtain the cooperation of the patient when a face mask is to be fitted or an anesthetic injection needs to be given. By holding the mask close to the patients face, the patient will inhale sufficient anesthetic gas to become sedated. From the 1950s various designs of top circuits have developed and classified from open breathing circuits, semi-open breathing circuits, semi-closed breathing circuits to closed breathing circuits. Well-known names for these circuits are Mapleson, Magill and Bain, who all contributed to the development of top circuits.

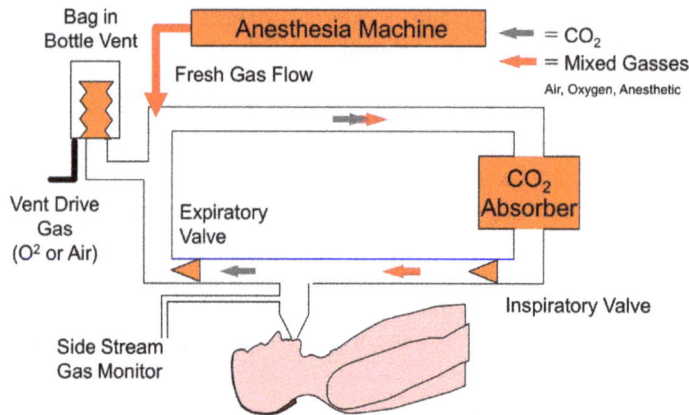

FIGURE 5.1.17   Circle system. "Low flow".

Figure 5.1.17 is fundamental to understanding how general anesthesia is delivered in today's OR. This simplified diagram shows how recycling the gasses dramatically reduces the amount of FGF required to a value of around 20%, and often even lower, compared with a top circuit. This represents a large saving on anesthetic vapors and anesthetics gasses such as nitrous oxide and helium. Not only is there financial savings to be made but also far less pollution and improved safety for all in the OR.

Let's look at the operation of the circle system from the point at which the sedated patient is connected via the ET tube to the anesthetic machine.

- • The anesthetist will first begin to deliver FGF by opening the various needle valves at the bottom of each flow tube in order to deliver the anesthetic mixed gasses to the circuit and the patient. It is usual at this point that this is just oxygen and can also be delivered by pressing the emergency oxygen bypass valve on the anesthetic machine. This flow then fills the circuit and the bellows within the 'Bag in Bottle Ventilator', which then rise to the top of the bottle. Any excess gas will lift a 'spill valve' which is positioned within the bellows chamber and an airway pressure limit valve (APL valve). Both valves allow the excess gas to be removed from the circuit to the scavenging system and out of the OR via a scavenging hose connected to a wall connector. These two valves are not shown in Figure 5.1.17. This procedure is often referred to as 'priming the circuit'.
- • The start of ventilation starts with 'Vent Drive Gas' being driven into the side of the bottle between the bellows and the glass bottle causing the bellows to compress downwards and

thus inflate the patient's lungs. The drive gas will stop at the end of the inspiration phase once the desired volume of gas has been delivered. This can be measured by reading the volume markings on the side of the bottle. Any excess pressure will be relieved by the APL valve. The patient then breaths out and the bellows return to the top of the bottle. Should there be any leak in the circuit, the bellows will fall short of reaching the top of the bottle and the anesthetist will then examine the circuit in order to find the source of the leak. In order to ensure the gas within the system flows only in one direction, there are two valves—the inspiratory valve and the expiratory valve. These valves can often be found on the gas block on the anesthesia machine, and can often be observed through glass covers opening and closing during ventilation.

- Crucial to the ability to recycle the gasses within the system is the carbon dioxide (CO_2) absorber. Usually placed close to the gas block, this cannister/jar is filled with soda lime crystals. As the CO_2, which has been produced by the patient, enters the absorber, a chemical reaction takes place that destroys the CO_2 and in doing so produces heat and water vapor. Both of these are beneficial to the patient as the FGF can be cool and dry, and so can be considered as a form of passive humidification. When using a soda lime absorber, the color changes as the soda lime becomes exhausted. Soda lime often starts as white and changes to purple when it becomes exhausted, but be aware that there are other color combinations.
- This is a closed system with a limited amount of gas, particularly oxygen within the circuit. It is therefore imperative that constant breath-by-breath analysis of the gasses within the system is undertaken. This is achieved by removing a small side-stream flow of the gas that is entering and leaving the patients lungs. This side-stream flow, often in the order of 100 mL/min is delivered to either the patient monitor or anesthetic machine for measurement and display. In doing so, the anesthetist is able to respond rapidly to the changes in the patient condition or such things as a gas leak in the circuit.

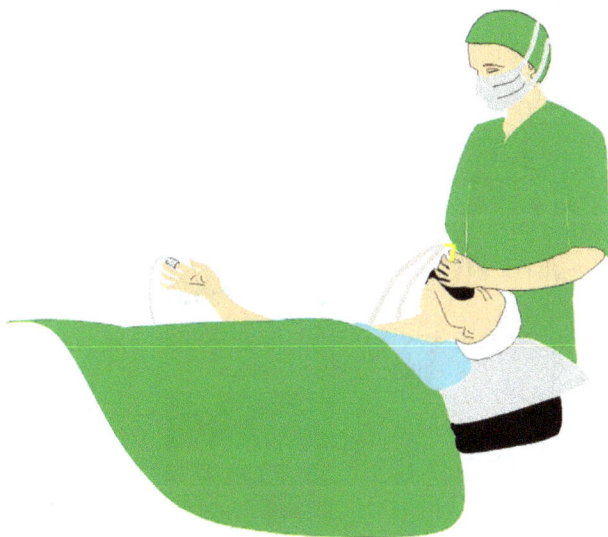

FIGURE 5.1.18 Arrival in the operating room/theatre.

For a conscious patient, the arrival in the OR will undoubtably be an anxious time (Figure 5.1.18). Although medical staff in the OR may see this as just routine, for the patient it may well be their first occasion in the OR and if they have had a previous general anesthetic and surgical procedure before they will reflect on their previous experience recalling the pain and sickly feelings while recovering. Many of us will react differently in such a situation and it is important that all the staff around the patient take into account the patient's emotional response, the need for privacy and individual respect. You will hopefully during your careers have occasion to be in the OR when a patient arrives for surgery, and so you should respond to their presence with understanding and respect for their privacy. It may be that on occasions you should leave the OR and return when the patient is no longer conscious. No one feels good about being told to leave by senior medical staff, so remain aware of the situation and 'read the room'.

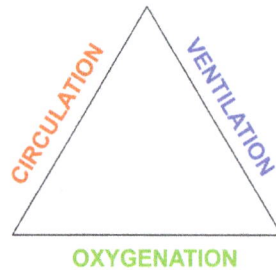

FIGURE 5.1.19   The need to "maintain".

At the very center of the role of the anesthetist is the need to maintain the three dynamics of circulation, ventilation and oxygenation (Figure 5.1.19). These three dynamics are greatly influenced by the anesthetic gasses, vapors, drugs, ventilators settings and not least the surgical procedure being undertaken. Constant monitoring is needed, not just of vital signs such as heart rate and blood pressure, but also such things as ventilator settings, the surgical procedure and fluid balance. These all feed into the decisions made by the anesthetist throughout the stages of induction, maintenance through the surgical procedure and the recovery phase in order that the patient safety is maintained.

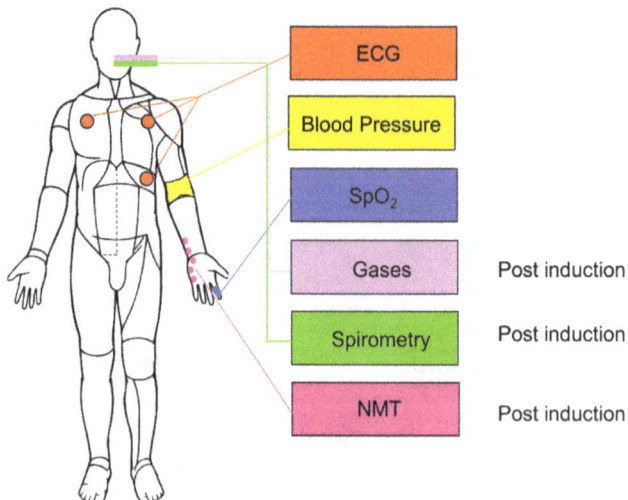

FIGURE 5.1.20   Preoperative monitoring.

If you have not done so at this point, it is important that you read the sections in Chapter 2 that detail the various monitoring parameters.

When the patient arrives in the OR one of the first things to be undertaken is that they are connected to the various monitoring parameters such as ECG, blood pressure and SpO_2 (Figure 5.1.20). The patient's initial parameter values while they are conscious and not sedated are recorded into the anesthetic record. Following induction, other initial parameter values are also noted in the anesthetic record. Airway gasses will provide information not just on level of oxygen provided (FiO_2) to the patient from the anesthetic machine, but also the oxygen uptake (FiO_2–FeO_2, fractional expired oxygen). The other anesthetic gasses and vapors will also be recorded into the anesthetic record. Spirometry, the measurement of gas flows, volumes and pressures are also recorded, and initial loops and values are useful in determining the changes within the lungs during the surgical procedure.

NMT is a measurement parameter undertaken by connecting electrodes usually to the wrist, palm and thumb along the path of the ulnar nerve in order to assess the effect of the muscle relaxant drugs that are administered to the patient in order to achieve paralysis. By providing an electrical stimulation signal to the ulnar nerve and measuring muscular transmission, the anesthetist is able to adjust the titration (delivery) of the muscle relaxant drug. There are a number of different methods used in measuring NMT such as EMG and AMG. One term often used with regard to this parameter is 'train-of-four' (TOF) monitoring which refers to four low voltage impulses being applied to the electrodes and accessing the thumb movement to each impulse. In short, the greater the effect of the muscle relaxant, the less the thumb will move.

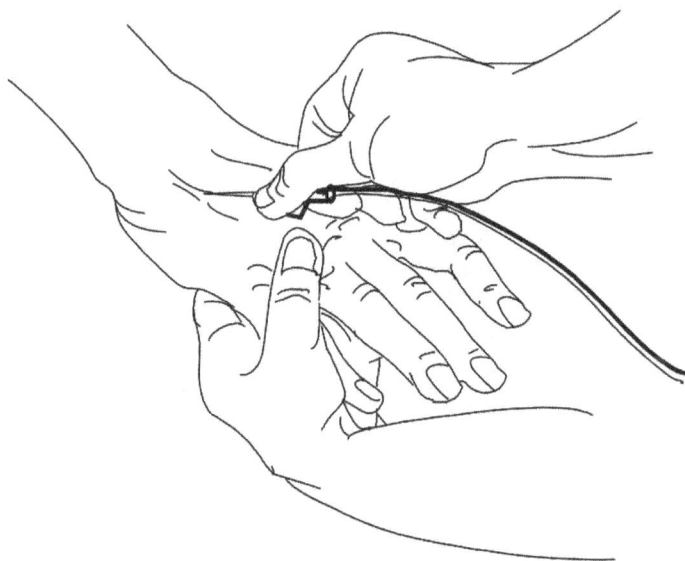

FIGURE 5.1.21   Intravenous cannulation.

An IV cannulation, sometimes referred to as a line is placed into a large vein in the back of the hand or arm (Figure 5.1.21). It often has a 'butterfly' mount that allows for some movement and is less painful for the patient. Through this 'line', the patient will have various drugs administered and it can also be used to extract blood samples.

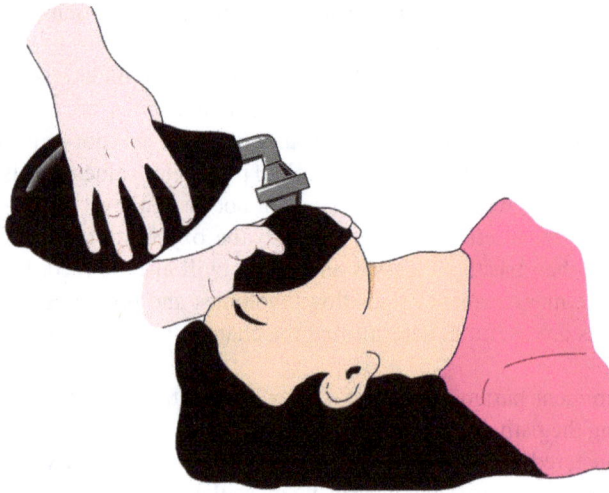

FIGURE 5.1.22   Preoxygenation.

One of the most important stages in the induction phase is to 'pre-oxygenate' the patient by administering 100% oxygen in order to build a reservoir within the patient's organs and tissues should there be a collapse of either ventilation or circulation (Figure 5.1.22). This reserve reservoir will allow more time for the anesthetist to urgently address the critical issue before the onset of hypoxia (lack of oxygen to the bodies, organs and tissues). Hypoxia is one of the most common errors that occurs in the OR and worldwide accounts for many incidents that result in brain damage, death and severe injury to the patient.

Induction Agents

- Sedation, Anti-nauseant, Drying Agent
- Diprivan (Propofol)
  - White and milky, large syringe
- Thiopentone
- Others are Epontol, Brietal, Ketamine, Etomidate

FIGURE 5.1.23

The drugs listed in Figure 5.1.23 are some of the most common drugs administered during the induction phase. Sedation, anti-nauseant and drying agent drugs are often given orally before the patient arrives in the OR. The anti-nauseant drugs, sometimes referred to as antiemetic, are given to alleviate the patients need to vomit. A drying agent drug is sometimes given in order to reduce the production of saliva, which would impede the airways during ventilation.

The purpose of listing many of the IV/intramuscular drugs is not to give a detailed account of their complex actions on the body, but rather to just allow the biomedical engineer a basic awareness when they hear these drugs being spoken about in the OR and what these drugs are being used for.

Diprivan and Thiopentone are two of the most commonly used drugs in the OR. They can be used for induction or throughout the surgical procedure. Given as an IV injection, they are very quick acting, often just seconds from the start of the injection to loss of consciousness and pain relief.

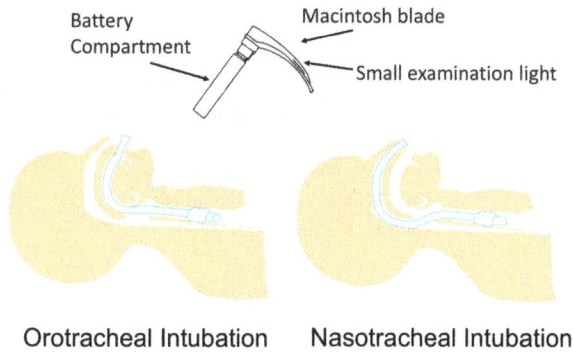

FIGURE 5.1.24   Endotracheal intubation.

Once the patient has been anesthetized and pre-oxygenated, the next procedure is to intubate the patient by fitting an ET tube (Figure 5.1.24). This is usually done with the assistance of a laryngo-scope which is a hand-held device with a blade that is inserted via the mouth toward the epiglottis in order for the anesthetist to be able to see the vocal cords and direct the ET tube through the vocal cords into the trachea. There are currently two types of blades used on the laryngoscope, the most commonly used blade for an adult is the Macintosh blade, which is curved and the Miller blade, which is straight, and often used for pediatric patients.

There is a great deal of skill required in order to insert and accurately place the ET tube into the trachea and the anesthetist will practice for many hours on manikins in order to become proficient at this important technique. Once the ET tube has been inserted, the anesthetist will check to ensure it is positioned in the trachea by connecting the ET tube to the ventilator and monitoring the airway gasses, particularly for the presence CO_2. Once the correct positioning of the ET tube is confirmed, the cuff will be inflated in order to ensure that airway gasses only pass through the center of the ET tube.

Guedal Stages

- **Clinical stages of Anesthesia**
  - Stage 1
    - Analgesia   beginning of induction to loss of consciousness ,"disorientation"
  - Stage 2
    - Excitement. Struggling, breath holding, coughing, swallowing

FIGURE 5.1.25

One method for the anesthetist to access the patient during anesthesia is to use the Guedel Stages assessment (Figure 5.1.25). As the patient begins to 'go-under', the anesthetic agents take effect; by looking at the patient's responses at this time allows the anesthetist to estimate when the patient is fully anesthetized and the surgical procedure can begin. Most patients will pass through the first two stages rapidly and without too much difficulty.

Guedal stages

- Third stage **surgical anesthesia**
  - Plane 1 automatic resp to no eye ball movement
  - Plane 2 to intercostal paralysis
  - Plane 3 full intercostal paralysis
  - Plane 4 diaphragm paralysis
- Fourth stage
  - Overdose Leading to death

FIGURE 5.1.26

Stage 3 is where the surgeon will prepare to start the surgical procedure (Figure 5.1.26). Opening the patient's eyelid and checking for lack of movement will indicate that Plane 1 has been reached. As the patients breathing becomes shallower and eventually leads to no chest movement, Planes 2 and 3 will have been attained. Finally in stage 3, Plane 4 will occur when the patient diaphragm no longer functions, and the patient ceases to spontaneously breath. At this point, the patient will now be dependent on the mechanical ventilator, or the bag, valve and mask/ET Tube connected (AMBU Bag) that is operated by hand by the anesthetist for their respiratory support. It is of course hoped that the depth of anesthesia induced will never reach Stage 4, excessive anesthetic agents could ultimately lead to the patient's demise.

- Sleep/consciousness
- Bleeding fluid balance
- Pain Analgesia
- Optimal relaxation muscle relaxants

FIGURE 5.1.27 Maintenance surgical procedure.

The critical factor is balance. The points in Figure 5.1.27 are constantly monitored and necessary adjustments made to the various anesthetic agents and drugs that control consciousness, pain relief analgesia and muscle relaxants. One other important factor that needs to be also kept is fluid balance. Fluids such as blood and urine are constantly being lost during medium/lengthy surgical procedures and are needed to be replaced almost continually during a medium/long surgical procedure. There are also added fluid losses that occur due to the heating effects of the OR room surgical lights that are used to direct bright penetrating light upon the surgical area for the surgeon. These surgical lights have the effect of causing evaporation not only from the patient's skin but also exposed internal organs.

Saline solutions bags hung from drip stands will use gravity to feed the solution into cannulas placed at various points on the patient's body such as the arms, legs and chest and contain such thing as vitamins, heparin, electrolytes, amino acids and drugs. Additional replacement blood is often administered using the same bag method. There are two other devices you will see being used to administer various solutions, these are syringe drivers and volumetric pumps. As a clinical engineer, you will undoubtably become involved in the servicing of these important devices at some point in your career.

Medical Gases

- Nitrous Oxide - N_2O - Blue
  - Discovered 1772 - used 1868
- "The Anesthetist's Friend". "The BULK agent".
  - Colourless gas. Non-irritant.
  - Very Environment Unfriendly (Ozone depleting)
- Not metabolised within the body. Rapid Uptake and Elimination. Can lead to Diffusion Hypoxia.
  - Low Potency, but good analgesic
  - Non-reactive with tubing or soda lime
  - Minimal impact on CVS and Respiratory Depression
- Entonox - White/Blue
  - 50% O^2 and 50% N^2O

FIGURE 5.1.28

A short recap, summarized in Figures 5.1.28 and 5.1.29: Possibly the first significant step in providing pain relief, as we know it today, was the discovery of nitrous oxide, sometimes known as 'laughing gas'. Joseph Priestley went on to demonstrate his discovery of nitrous oxide in 1772, but it wasn't until a dentist named Horace Wells demonstrated it as an anesthetic agent during a dental extraction in December 1844, and that its medical properties were accepted and recognized as a significant anesthetic agent.

For more than 100 years, nitrous oxide has been an important gas used in anesthesia, offering both sedation and pain relief (analgesia). It has many positive properties, such as being a non-irritant and inexpensive, but there are also a number of major drawbacks. One of the most important drawbacks is that it is very environmentally damaging, not just within the OR but also when it is expelled to the outside environment where it will react with sun light and cause ozone depletion of the upper atmosphere in a similar way to the effect of refrigerant CFCs (chlorofluorocarbons).

One of the characteristics of nitrous oxide is that it is not metabolized (chemically broken down into other chemical compounds) within the body giving it the property of rapid uptake into the blood stream and rapid elimination by means of being expelled with breathing.

Please keep in mind that although nitrous oxide may sound that it would contain oxygen, in fact it does not contain any oxygen usable by the patient. Oxygen molecules contain two oxygen atoms, whereas N_2O only contains one oxygen atom. As N_2O is heavier than air, there is a risk that if there is insufficient oxygen provided along with the N_2O, the alveoli will be saturated with N_2O and thus unable to absorb enough oxygen to meet the patient's needs. This situation is referred to as diffusion hypoxia and can be extremely dangerous to the patient. Today the use of nitrous oxide has been greatly reduced for the reasons above and many hospitals are removing nitrous oxide pipelines. Very often in anesthesia, air and oxygen are solely used with the anesthesia being a combination of vapors and IV anesthetic drugs.

Entonox, often referred to as 'gas and air' is still used widely worldwide, particularly for the later stages during labor for mothers-to-be. A typical technique used is to allow the mother to hold a face mask or mouthpiece to the mouth and nose and inhale the mixture of 50% nitrous oxide and 50% oxygen. There is an in-line valve that only opens to supply gas when the patient draws in gas during inspiration. This would provide a high degree of pain relief from labor pains. However, recently there has been a raised awareness that the contaminated environment around the mother-to-be would expose medical staff such as midwives to unnecessary exposure of nitrous oxide regularly and the long-term health consequences for medical staff might be detrimental. In many hospitals, Entonox has now been withdrawn for this reason.

Medical Gases

- Heliox
  Is a mixture of oxygen and helium that has
  low gas density (lighter that air).
- Nitric Oxide - NO
  - Often used during induction for paediatric
    patients.

FIGURE 5.1.29

Heliox (Heliox21, contains 21% oxygen and 79% helium) is occasionally used in anesthesia to aid in the delivery of vapors/volatile and oxygen for patients that may have respiratory conditions such as ARDs, asthma or COPD. Being lighter than air and oxygen, it has the capability to deliver the vapors deeper into the lungs and thus improve the uptake of the anesthetic vapor and oxygen.

Nitric oxide (NO—not to be confused with nitrous oxide N_2O) is commonly used as an induction agent for pediatrics and is administered by holding a face mask close to the patient's nose and mouth. Sedation happens very quickly allowing the anesthetist to continue with other standard anesthetic agents. Nitric oxide can now be found being used in the intensive therapy units and neonatal intensive care units and is delivered by specialized mechanical ventilators.

- Method by which the volatile
  agents can be vaporised within
  the F.G.F.
- At a calibrated dosage
  - Low Accuracy
  - Fitted to the Anesthesia
    Machines "Back Bar"
  - Only ONE used at a time.
    Interlock system
  - Handle with GREAT CARE

FIGURE 5.1.30   Vaporizers.

Today's vaporizers are highly complex devices that are heavy and require very careful handling (Figure 5.1.30). The first important point for the clinical engineer to be aware of is that you should not seek to remove a vaporizer from an anesthesia machine without first ensuring that the liquid vapor has been fully drained and removed by authorized clinical staff. The safest practice is to always have the vaporizers removed by clinical staff before attempting to work on anesthesia machines. If the liquid anesthetic agent were to leak out, then you and others could be exposed to a dangerous amount of anesthetic vapor causing loss of consciousness, injury and even worse.

The accuracy today of vaporizers have greatly improved percentage accuracy, but is still not absolute, so it is usual for the anesthetist to rely on the reading from the patient gas monitoring system when setting the percentage of vapor being delivered as this offers the best degree of accuracy. Within the vaporizer are controls that adjust for gas temperature and flow. At the front of the vaporizer is a fluid level indicator, fill/empty port and on the top a rotational control to set the percentage delivered. At the back of a vaporizer are two connection ports that will sit on two connectors mounted on a back bar. These are the input and output connectors for the FGF. Within the back bar is a mechanical inter-lock that working with the vaporizers ensure that only one vaporizer can be operated at one time. This is to ensure that it is not possible to double-dose the patient with vapor.

FIGURE 5.1.31   Volatile/vapors anesthetics.

Listed in Figure 5.1.31 are the five most common volatile/vapors. One term that might need explaining is the term 'volatile'. This term was first used in anesthesia when ether was commonly used and referred to the very real danger that a spark or flame would cause the vapor to ignite in the same way gasoline fumes do. As a liquid, the anesthetic agents listed above no longer present this risk and will not ignite, but the term to describe them in their liquid form has remained volatile.

Halothane: Discovered in 1951 and used 1958. This was the most popular of today's agents but has recently been withdrawn due to the adverse side effects. It is a halogenated ether, good hypnotic properties, some analgesic affect, no real muscle relaxant effect. It is a colorless liquid that decomposes on exposure to light, so is stored in amber bottles. More soluble in blood than enflurane or isoflurane, so you can expect a slower induction and recovery. It shows effect on both CNS (central nervous system) and CVS (cardiovascular system).

Enflurane: Discovered in 1965 and used 1973. Halogenated ether but because of relative low blood/gas solubility, induction is rapid and depth of anesthesia is readily altered. Prompt recovery usually follows, depending on when the vapor was turned off. Good hypnotic properties, some analgesic effect, some muscle relaxant effect but does affect CNS and CVS.

Isoflurane: Discovered in 1965 and used 1972. It is similar to enflurane and shares many properties with it. Like enflurane, it is fluorinated and chlorinated, which is essential for nonflammability. It is stable and there is no need for stabilizers, not affected by alkali or UV light but has a pungent smell. Induction with isoflurane will be more rapid than the others. The low solubility will also mean prompt awakening but is still dependent (as are the other agents) on dose-time relationships. A 'blow-off' period is still essential even if not as long as for other agents.

Desflurane: Until recently desflurane was widely used but because of it environmental damaging properties it has now been banned in a number of countries. A special designed tec 6 heated vaporizer had to be used to ensure the stability of the concentration as any small change in temperature would result in large changes in vapor concentration. It was considered a good day case agent as the patient recovered rapidly from its effects.

Sevoflurane is extensively used today as it is less damaging to the environment and offers similar anesthetic properties to desflurane. It is good for induction and has a rapid recovery time. The cost of it has over recent years reduced and so its popularity has increased. As with all anesthetic volatile agents there are side effects. Nausea or vomiting are the most common but also cardiac arrhythmias, confusion are just some of the side effects experienced by patients.

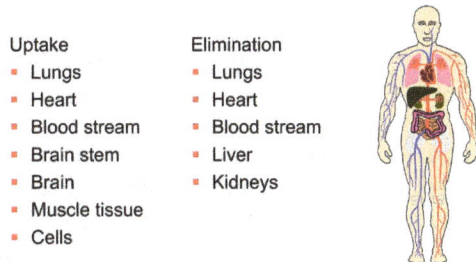

FIGURE 5.1.32   Uptake and elimination of inhalation agents.

In knowing how the body absorbs and eliminates inhalational anesthetic agents, it is important to be able to understand the major impact that anesthetic agents have on such organs as the brain, heart, liver and kidneys. These will all to some degree metabolize the agents and in doing so are profoundly negatively impacted by what could be thought normally as toxins. The chemical breaking down of these agents by the various organs produce a mixture of chemical compounds and gasses which are then transported by the blood stream to the lungs to be breathed out. The kidneys and liver will then process any remaining anesthetic agents for excretion in urine and faces, thus providing the vital role of elimination (Figure 5.1.32). Once the anesthetic inhalational agents are removed from the patient, the various organs will return to normal functioning given some time, without long-term side effects in most cases. A similar process of uptake and elimination is also undertaken when considering IV and intramuscular anesthesia, which very much involves the kidneys and liver.

Minimum alveolar concentration

- **MAC** is a concept used to compare the strengths of anesthetic gasses and vapours.
  - Each agent has a differing potency for a given concentration
- MAC 1 = 50% of subjects show no response to surgical (pain) stimulus.
- MAC 1.3 = would block response in 95% of subjects.

FIGURE 5.1.33

You may notice when viewing the screens on the patient monitor or anesthetic machine that next to the values for gasses and agents that there is a value given for minimum alveolar concentration (MAC). MAC is a means of anticipating the depth of anesthesia that will occur in a patient when administered with the various anesthetic gasses and vapors (Figure 5.1.33). You may think of this as an anesthetic strength or potency. MAC 1 is where 50% of all adult patients will show no response to surgical incision. For example, MAC 1 for sevoflurane is usually a setting of 2.31%–2.84% with 100% oxygen. Each anesthetic vapor and gas has an individual percentage value that corresponds to MAC 1. This enables the anesthetist to readily anticipate the depth of anesthesia when either giving a single anesthetic agent or a combination of agents and gasses.

MAC is not a measurement of the patient's output gasses and vapors, but a measurement and calculation of the gasses and vapors delivered by the anesthetic machine to the patient. The process of measurement of the MAC value is today normally calculated in real-time by the patient monitor or anesthetic machine. Taking a side stream sample of gas from the FGF at the top of the patient's ET tube and drawing it into the patient monitor or anesthetic machine uses various internal gas analyzers to calculate the percentage concentration of each agent and/or gasses in the FGF. Using a software table for each of the various agents/gasses, it is possible that the individual and combined MAC values can be calculated and displayed on the patient monitor or anesthesia machine.

One of the most important developments in anesthesia over the last three decades has been the development of advance depth of anesthesia monitoring. BIS and E-entropy have greatly improved the ability of the anesthetist to access the effects of the anesthesia for individual patients. By monitoring electrical activity within the brain, it is possible to access the depth of anesthesia of the patient. BIS, for example, will display a reading of 100 for a completely awake patient, down to a reading of ~40 for a patient that is deeply anesthetized.

	EXAMPLES
Anesthetics	Inhalational anesthetics Intravenous anesthetics - propofol - etomidate - ketamine
Analgesics	- fentanyl - alfentanyl - sufentanil Potentiate the anesthetic effects of intravenous and inhaled anesthetic agents
Muscle relaxants	- succinyl choline - atracurium - rocuronium - pancuronium - vecuronium Potentiated by inhaled anesthetic agents

FIGURE 5.1.34 Drugs during anesthesia.

This book does not describe in detail the chemical composition and physiological effects on the patient of each of these drugs, as clinical engineers will not be expected to participate in discussions with clinicians on such matters. The list of drugs in Figure 5.1.34 only seeks to demonstrate why certain drugs are selected for use such as analgesics (pain relief), muscle relaxants and consciousness in order for a clinical engineer to have some understanding when they encounter an anesthetist using them. One thing that you may have noticed on the list of muscle relaxant drugs is that the last four have within their names 'coronium', which indicates that they are all originally derived from 'curare' that was used in South America on the tips of arrows to paralyze prey when hunting.

All the drugs listed above are 'controlled drugs', meaning that they are only approved for use by clinicians such as anesthetists in certain circumstances and are stored under lock and key for security. When they are removed from the drug cabinets in which they are stored, the person removing them will record details of the amount taken from the cabinet, patient details, date, time along with the signatures of those removing the drugs. Often this requires the signature of two clinicians in order to protect against mistakes or malpractice.

I.V. Analgesics Agents

- Morphine first commercially sold in 1827
- Diamorphine – Heroin is an opioid drug synthesized from morphine, which is a derivative of the opium poppy
- Propofol (marketed as Diprivan by AstraZeneca)
- Pethidine was the first synthetic opioid synthesized in 1932
- Fentanyl is approximately 100 times more potent than morphine

FIGURE 5.1.35

Figure 5.1.35 gives a little more detail regarding some of the IV analgesics agents that you may hear discussed in the OR. Morphine is well known and often referred to in relation to emergency pain relief in such places as battlefields, when given by army medics to wounded soldiers. Diamorphine is commonly known as heroin and is often crudely distilled from the opium poppy. In the hospital environment, it is known as diamorphine that has a higher-grade quality and consistency. Propofol/Diprivan is very recognizable as it is a white milky solution in a large hypodermic needle. Pethidine is possibly the most commonly used IV analgesic agent used today for acute pain relief. Fentanyl as you can see from the slide is an exceptionally powerful analgesic agent and is administered with great care. Over the recent years you have heard of the 'opioid crisis' that effects many western countries such as the United States. This is where many of the drugs listed above, which are all addictive, have become widely available by either illegal or legal channel to the general population. A tendency by some physicians to overprescribe these powerful drugs has led to addiction and sometimes death of many patients.

I.V. and I.M. Muscle Relaxants
- Depolarizing and None - depolarizing
  - Long acting and short acting blockade,
  - Antidote, mechanism of reversal of a muscle blockade.
- Suxamethonium
- Curare
  - Alcuronium
  - Atracurium
  - Pancuronium
  - Vecuronium

FIGURE 5.1.36

Suxamethonium is a short acting depolarizing (effectively overwhelms the connection between nerve endings and muscle tissue) muscle relaxant. It usually provides just a few minutes of muscle paralysis in order for the patient to be intubated. The curare-based muscle relaxants act differently by coating the nerve ending connection to the muscle and thus preventing the transmission of the electro-chemical signal from the nerve to reaching the muscle. These curare-based muscle relaxants have a far longer effect and are administered throughout the surgical procedure in order to maintain muscle paralysis (Figure 5.1.36). In an emergency there may be a requirement to quickly reverse the effect of a muscle relaxant, so an antidote drug (a drug that neutralizes the effects) will be administered.

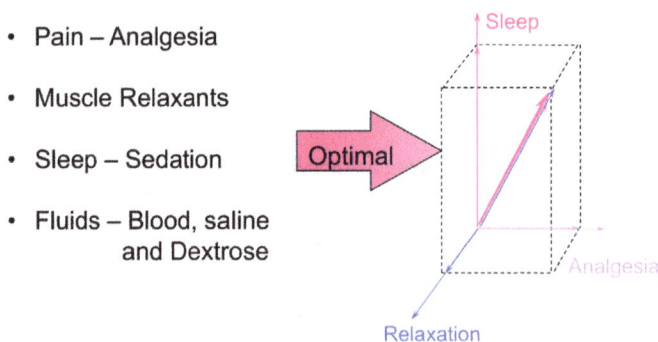

- Pain – Analgesia
- Muscle Relaxants
- Sleep – Sedation
- Fluids – Blood, saline and Dextrose

FIGURE 5.1.37   Continual monitoring and maintenance throughout the surgical procedure.

Throughout the surgical procedure the anesthetist will constantly monitor and adjust not just the three parameters of pain, sedation and muscle relaxation, but also fluids (Figure 5.1.37). Adjustments of anesthetic agents that regulate pain relief and sedation, along with administration of muscle relaxants to maintain flaccid muscles in the surgical area is continually undertaken.

It is inevitable that some fluids will be lost during surgery due to incisions and evaporation and these will need to be replaced in order to maintain hemostasis of the patient. You will inevitably see 'drips', clear plastic bags hung from steel poles containing blood and other fluids such as saline (a mixture of water and salt, sodium chloride) and dextrose (a sugar solution) being administered via IV lines into the patient.

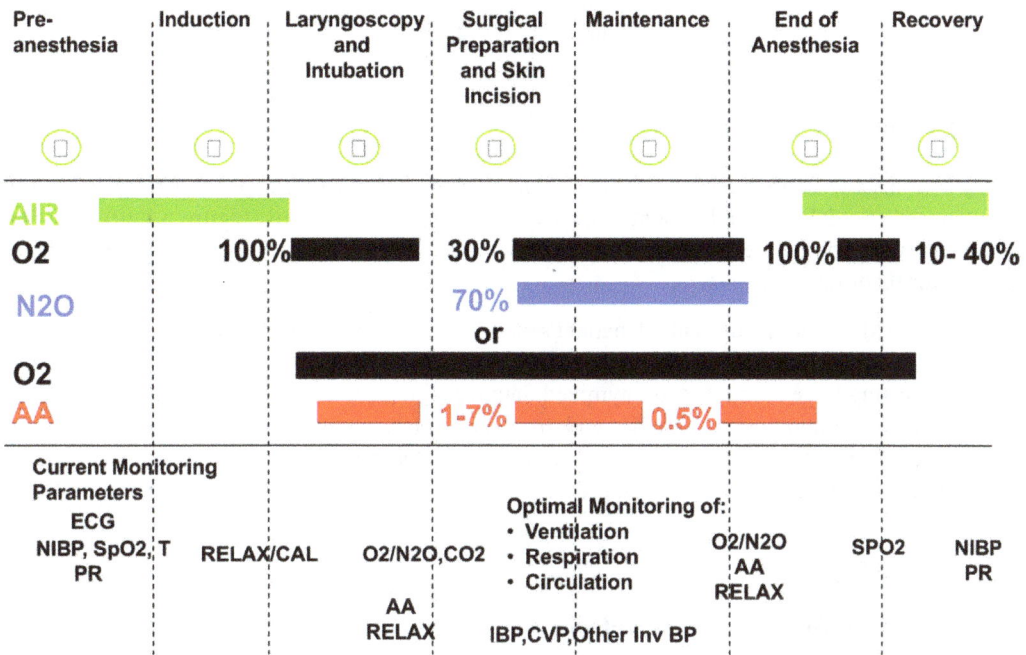

FIGURE 5.1.38   Phases of anesthesia.

Figure 5.1.38 is designed to convey the various levels of anesthetic gasses, vapors, air, oxygen and monitoring and the various stages a patient will pass through during their time in the OR. It is not a definitive representation, only a basic guide as to what might be occurring at any particular time. If you get the opportunity to attend the OR during a surgical procedure, you may find it useful to undertake with the assistance of the anesthetic staff drawing your own chart as drawn above in order to gain a deeper understanding of the anesthetic procedure and the monitoring requirements. As previously stated, the use of N_2O is very much on the decline, but there are some countries where it is still in use.

End of Anesthesia

Restoration of spontaneous

•Breathing

•Circulation

•Extubation

•Awareness

FIGURE 5.1.39

The recovery phase of anesthesia is of course one of the most important stages that will need to be fully completed before the patient's surgical procedure can be said to be a success (Figure 5.1.39). This often occurs in a specialist area within the operating rooms often known as 'Recovery'. Here specialist nurses attend to the patient as they recover not only from the surgical procedure, but also the effects of the anesthetic. It is expected that during this recovery period that the anesthetist will call in to check on the patient's progress before allowing them to either return to the ward or being discharged from the hospital.

- Patient demographics
- Pre-op information
- Drugs
- Fluids
- Vital signs
- Events

FIGURE 5.1.40   Patient data management.

Today's modern ORs are very much 'data connected' with almost all aspects of surgical procedures, anesthetic delivery and patient response being digitally recorded (Figure 5.1.40). This has in many developed counties become a mandatory requirement. There are basically three primary reasons for recording this data.

- The need to record individual patient's data should there be any question about what occurred in the OR during the surgery.
- Harvesting large amount of anonymized data for clinical research and development taken by such institutions as universities and drug companies.
- Hospital operational planning to ensure the most effective use of resources such as staff, equipment and facilities, for example, operating rooms.

SECTION 5.2

ANESTHESIA MACHINE PRINCIPLES

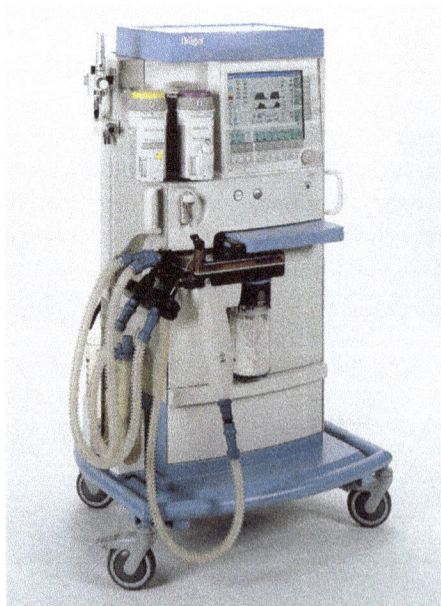

FIGURE 5.2.1   The anesthetic machine – purpose and function.

Before studying this section on anesthetic machines, it is essential to study the section on anesthesia principles and terminology to better understand the process. This section on anesthesia machines is only intended to provide some information of the standard components, their functions and basic principles. It is definitely not sufficiently detailed enough to provide enough knowledge for a clinical engineer to work safely on any anesthesia machine. In order for a clinical engineer to work on an anesthesia machine, it is essential that they undertake formal training and certification, either from the machine manufacturer/supplier or from the hospital/organization's own training department. What is offered here should enable the clinical engineer to identify and understand the functions of many of the mechanical and pneumatic standard components that go into a basic anesthesia machine (Figure 5.2.1). Over recent decades many manufacturers have developed highly sophisticated machines that incorporate complex electronic and software systems that provide a greater range of gas/vapor delivery and mechanical ventilation and this section will not attempt to explain these advanced machines.

What the anesthetic machine does

- The <u>safe</u> delivery of:
  - Oxygen
  - Anesthetic Gasses
  - Anesthetic Agents
  - If incorporated – Mechanical ventilation
  - If incorporated – monitoring of either/both machine delivery, patient expelled gasses

FIGURE 5.2.2

Figure 5.2.2 lists the five primary functions of a basic anesthetic machine. The delivery of oxygen is the single most important function of the machine, and the design of the machine always ensures that oxygen can be delivered, particularly when needed quickly in an emergency. You may think of the anesthetic gasses, anesthetic agents/vapors and oxygen as being blended together within the machine and leaving via the common gas outlet (CGO) to the circle or top circuit, and thus to the patient. Today's anesthesia machines all incorporate an integrated mechanical/electronic ventilator, but there are still some machines in use that have a separate ventilator. Today's advanced anesthesia machines incorporate monitoring systems that not only monitor patient expelled gasses, flows, volumes and pressures, but just as importantly display the machines setting of the ventilator, gasses (electronic flow meter readings), vaporizers and wall/cylinder gas pressures to ensure maximum patient safety.

Primary Components of an Anesthetic Machine

- Gas supplies: From the central pipeline to the machine as well as cylinders.
- Flow meters.
- Vaporizers (when fitted).
- Fresh gas delivery: Breathing systems and ventilators.
- Scavenging.
- Monitoring.

FIGURE 5.2.3

Figure 5.2.3 shows again the components that will be found in any modern anesthesia machine. One component that you may not have been aware of is the scavenging system. The scavenging system plays a vital role in protecting all the individuals in the OR from exposure to anesthetic gasses/vapors. It draws away excess gasses/vapors not only from the patient circuit but also via a small open port at the bottom of the anesthesia machine where any gases/vapors that may have spilled accidentally from the patient circuit. Fortunately, most gasses and vapors are heavier than air and collect on the floor of the OR. A low negative pressure is supplied via a wall connection (scavenging), and this draws the excess gasses/vapors from both the patient circuit and the small open port at the bottom of the machine. Eventually the excess gasses/vapors are released to the atmosphere via an open port on the roof of the building. However, today many hospitals have systems installed that can collect and neutralize these environmentally damaging gasses.

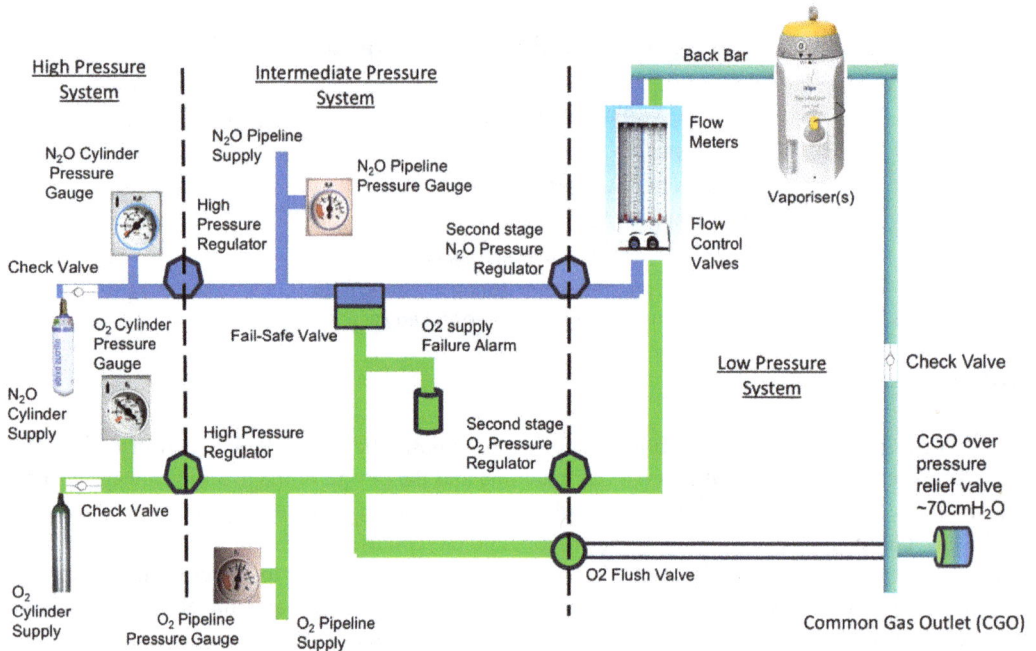

FIGURE 5.2.4   A basic anesthetic machine (without medical air supply).

Figure 5.2.4 shows a highly simplified diagram of an anesthetic machine (without a medical air supply) and shows how the machine can be divided into three pressure systems. The high-pressure system is concerned with the inputs of gasses from the cylinders attached to the back of the machine. The pressures within an 'E' size cylinder are extremely high and are regulated down to approximately 3.5 bars (50 psi) before entering the medium pressure system.

HIGH PRESSURE SYSTEM

Starting on the left side of Figure 5.2.4, we can see the two emergency supply cylinders: oxygen and nitrous oxide. Both are attached to the machine by a cylinder yoke that contains a one-way check valve to ensure that gas does not escape when the cylinders are either closed or removed. Next are cylinder pressure gauges that are displayed on the front of the machine and indicate contents volume in each cylinder. Following are the two high to low pressure regulators that reduce the pressures from 3.5 bar and 3.8 bar. This means that the 4 bar pipeline pressure that is supplied by hoses connected to the back of the machine from the medical gas supplies located on the OR walls would normally be the operating supplies. Only in the event of a wall gas failure will the cylinders be opened to provide supply to the machine.

To ensure security of supply, each gas pipeline from the wall has a pressure gauge indicator on the front of the machine. The cylinders attached to the back of the machine should always be considered as backup to the wall gas supplies which at 4 bar (58 psi) normally supply the anesthetic machine when it is in use. One of the daily checks is to ensure that the cylinder regulators, which are on top of the cylinders and are closed when the anesthetic machine is in use, are opened for a short period of time. The anesthetic staff will ensure that the cylinder pressure gauges on the front of the machine indicate full cylinders for each of the gas cylinders connected. It is very important that each cylinder is then switched off and that the anesthetic machine returns to being supplied from the wall gas supply only.

INTERMEDIATE PRESSURE SYSTEM

Within the intermediate pressure system are two very important components that provide patient safety if the oxygen supply fails. First is the fail-safe valve that only allows N_2O to flow toward the CGO if O_2 pressure remains correct. If the O_2 supply fails, then this spring-loaded valve will close the supply of N_2O and thus ensure that the patient does not receive 100% N_2O. The second is the O_2 supply failure alarm, which consists of a small tank containing O_2. With the O_2 supply being normal, a valve at the top of the tank is forced closed and ensures the O_2 within the tank remains.

Should the intermediate O_2 pressure fail, a valve within the tank opens and allows the tank's O_2 to escape to the atmosphere through a small whistle. In the event of a failure in the wall oxygen gas supply, the anesthetist will hear an oxygen failure alarm whistle and will react by turning on the oxygen cylinders at the back of the anesthetic machine until the supply to the anesthetic machine wall gas supply is restored. The audible alarm is around 55 dba and lasts for approximately 7 seconds. The anesthetist should always ensure that spare cylinders are readily available should the wall gas supplies not be restored during the surgical procedure.

Should the back-up oxygen cylinders not be turned off following the daily check, the following will then happen: the 4 bar oxygen wall gas supply being 0.5 bar higher than the 3.5 bar regulated cylinder supply ensures that a fail-safe valve situated between the cylinder gas supply and the wall gas supply remains firmly closed and does not allow any gas from the cylinders to enter the intermediate pressure system. If the wall gas supply were to fail, then the cylinder 3.5 bar supply will force open the fail-safe valve and supply the intermediate pressure system. The problem with this situation is that the anesthetist may not be made aware of oxygen wall gas supply failing, and will only be notified by the oxygen supply failure alarm whistle when the cylinder becomes depleted. If a spare cylinder is not readily available, the patient could suffer hypoxia. The only indication of the wall oxygen supply failure will be that the oxygen wall gas gauge will fall to zero.

LOW PRESSURE SYSTEM

For each gas there are second stage regulators that further reduce the pressures to ~0.5 bar. The outputs of these two regulators then flow to the inlets at the bottom of the flow control valves that are used to adjust the flows through the flow meters in order to control the gasses to the patient from the CGO. The flow control valves are essentially needle valves that are controlled by the anesthetist by reading the flow levels shown by the flow meters. A more detailed explanation will follow in this chapter of both the needle valves and 'flow meters'. On leaving the flow meter block, the mixed gasses now enter the back bar upon which the vaporizer(s) are mounted. When a vaporizer is opened then all the flow will enter the vaporizer and the flow will then be divided internally within the vaporizer with some of the FGF being diverted into the chamber containing the liquid agent. The volume of flow through this chamber is dependent on the concentration setting set by the anesthetist on the top of the vaporizer. The FGF then returns to the back bar and passes through a one-way check valve. This check valve ensures that there can be no back flow from the patient circuit into the anesthetic machine. For additional safety, a fixed CGO over pressure relief valve is mounted close to the CGO to ensure that the patient is protected from excess patient circuit pressure above ~70 dmH$_2$O.

Pressure in a standard 'E' size cylinder is 137bar

Cylinders and their yoke assemblies

Cylinder colour coding and PIN index systems

Always replace/check 'Bodok Cylinder Seals'

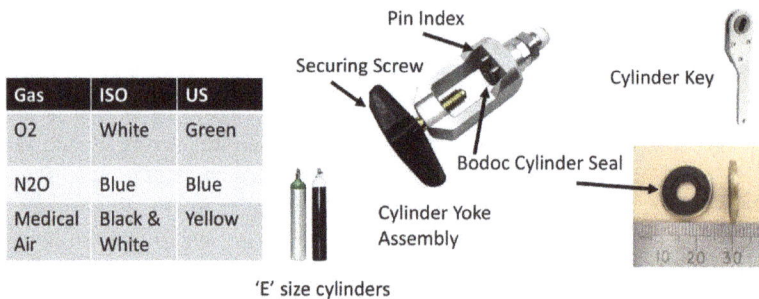

Gas	ISO	US
O2	White	Green
N2O	Blue	Blue
Medical Air	Black & White	Yellow

'E' size cylinders

FIGURE 5.2.5   High pressure system.

Pressure in a standard 'E' size cylinder is 137 bar (~2000 psi), which is extremely high and therefore requires great care when handling (Figure 5.2.5). When transporting any large medical gas cylinder, it is important that safety precautions are always followed. Always use gas cylinder trolleys and secure the cylinder before moving. Ensure that, if you are moving the cylinder any distance, people on the path you are taking are aware of what you are doing. If the gas cylinder proves too heavy to be lifted by one individual, then seek the assistance of another person as you mount the cylinder in the cylinder yoke.

It is advisable that every time a cylinder is changed, the Bodoc cylinder seal is also changed. It is good practice for any engineer working in the OR to always carry spare Bodoc seals and a spare cylinder key. When working with any anesthetic machine, you should always ensure that a cylinder key is attached by a chain to the rear of the machine or is readily available. Some cylinder manufactures provide a control knob upon the top of the cylinder in place of needing a cylinder key.

When fitting a replacement cylinder always ensure that you are fitting the correct gas cylinder to the correct gas yoke. Within the yoke there two pin index pins that are positioned differently for each particular medical gas. These two pins will locate into two holes within the cylinder head. Should the wrong cylinder be attempted to be mounted into the wrong yoke, the pin index pins will not align with the hole in the cylinder head and in doing so will not allow the securing screw to tighten the cylinder head flush against the cylinder yoke.

Gas cylinder colors—Currently there are two international standards used for the identification of gas cylinders. The International Standards Organization (ISO), and the US standard. Some cylinder top colors are the same such as nitrous oxide (Blue). Others such as oxygen (ISO: White; US: Green) and medical air (ISO: Black and White; US: Yellow) vary and it is important that you make yourself aware which gas color standard your hospital or organization use. I would strongly advise that, if possible, you attend a medical gas training course to ensure that you are fully familiar with the safe use, storage and handling of medical gas cylinders.

INTERMEDIATE PRESSURE SYSTEM

The intermediate pressure system consists of:

1. Pipeline inlet connections
2. Master switch (present in newer machines)
3. Pipeline pressure indicators
4. Second stage pressure regulators
5. Auxiliary gas outlets for ventilators
6. Oxygen pressure failure devices
7. Oxygen flush and the flow control valves

FIGURE 5.2.6

Figure 5.2.6 shows most of the components you have seen in the basic anesthetic machine drawing previously. There are however a couple of additions that you may find on a modern anesthetic machines. The master switch refers to a toggle on the machine that starts the electronic control systems and monitoring. In many anesthetic machines today, it is possible to continue to use the machine in the event of a mains supply failure as they have a battery backup system within the machine. Many modern anesthetic machines also provide an auxiliary gas outlet to drive external ventilators via a mini Schraider port. This port can also be used to provide via an external flow meter supplemental oxygen that is often used during the patient induction phase.

The probe for each gas supply has a protruding indexing collar with a unique diameter, which fits the Schrader socket assembly for the same gas only

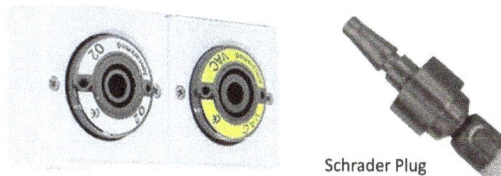

Schrader Plug

FIGURE 5.2.7   Schrader probe.

The Schrader medical gas connection system is possibly the most common medical gas connection system today, but it is not the only medical gas connection system in use. The Schrader probes are designed with an indexing collar and diameter that ensures that only the correct gas hose will fit the correct gas outlet. The probe is fitted by forcing the probe into the outlet and then apply a small clockwise twist to the probe that then is locked into position (Figure 5.2.7). This is a very similar action to fitting a 'bayonet' fitting light bulb into a bayonet light socket. To release the Schrader probe from the socket, you are required to press the outer ring on the socket. Please be aware that there are two sizes of Schrader probes and sockets, standard and mini.

- NIST - Non-Interchangeable Screw Thread
- DISS - Diameter Index Safety System
- Hoses are flexible, colour-coded and have built in reinforcements in the wall to make them kink proof.
- Note: All Gas hoses and connectors are routinely tested and certified by Medical Engineering

FIGURE 5.2.8   NIST, DISS and hoses.

An alternative to the Schrader connection system is non-interchangeable screw thread (NIST) and diameter index safety system (DISS). Both of these connectors ensure that it is impossible to connect the wrong hose to the wrong wall gas connection (Figure 5.2.8). With NIST, although the diameter of the connectors is the same for each gas, due to the different screw thread cut in both the male and female parts, it is impossible to interchange the wrong hoses. With the DISS connection system, due to each medical gas having a different diameter of probe and socket, again it is impossible to connect the wrong probes. When replacing hoses, it is always advisable to check carefully that the correct color hoses with the correct connectors are replaced as it has been known for manufactures to supply incorrectly color-coded hoses and connectors.

Medical gas hoses are often supplied by specialist medical gas hose manufactures, but sometimes are produced on-site within the hospital medical engineering departments. In many countries where this happens, the hospital may be required to register with the countries relevant authorities in order to obtain permission/certification to do so. Strict quality standards covering manufacturing, testing and verification have to be adhered to. Medical gas hoses are often routinely changed for new hoses after a given period of service due to the high-pressure gases they carry and to reduce the risk of failure due to fatigue. They are also regularly pressure tested and inspected throughout their period of service to ensure maximum safety.

- On almost all anesthetic machines both Wall Pipeline and Cylinder Gauges are provided.
- Wall Pipeline gauges indicate 'Pressure' ~ 4bar
- Cylinder gauges indicate 'Pressure' that directly relates to 'Volume' left within the cylinder

FIGURE 5.2.9   Wall and gas cylinder gauges.

One of the most important checks undertaken in the OR is to read the pressures on all the gas supplies. This is routinely done not just at the start of the working day, but also regularly throughout the day. One common error that can occur is to 'see' but not 'read', which means that due to a lack of concentration it is possible for the anesthetist not to be fully aware of the drop in pressure of a gas pipeline feed or leak within the cylinder supply system (i.e., leaking Bodoc seal). In some of the most modern anesthetic machines, there are second display and alarm systems that continually monitor these pressures.

The flow control knob for oxygen is the largest, most protruding and has tactile differentiating features like a fluted profile for additional and easy identification.

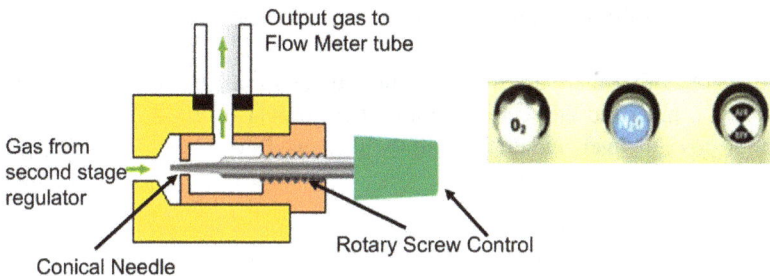

FIGURE 5.2.10   Rotameters (needle waves).

Rotameters is the name given to the rotary control needle valves (Figure 5.2.10) used to set the flow of each medical gas that the anesthetic machine supplies to the CGO. The conical needle varies the amount of flow through the valve as set by the rotary screw control. These precision valves are routinely cleaned and serviced due to contaminate particles from the wall/cylinder gas supplies not being totally clean. There is also the possibility that if the valve is over-tightened then damage will occur to the conical needle, and this can produce problems in the accurate control of the gas flow. For oxygen, the control knob has a distinctly different shape and diameter in order that without looking the anesthetist is able to differentiate oxygen from the other gasses even in dark lighting conditions.

- Conical in shape allowing for non-linear scaling
- The tubes have an antistatic coating on both surfaces, preventing the bobbin from sticking.
- Tubes, which measure flow, have different lengths and diameters. Some machines have a pin-index system at each end.

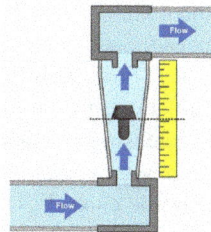

FIGURE 5.2.11   Flow tubes.

The amount of each individual gas flow going toward the CGO, via the vaporizer, is shown using glass/perspex, nonlinear transparent tubes that are marked vertically with the flow speed clearly printed on the tube (Figure 5.2.11). The internal bobbin will rise in accordance with the level of flow. There are two types of bobbins that can be found in flow tubes, the most common is a conical bobbin that should be read from the glass/Perspex's printed scale from the top of the bobbin. The second is a ball bobbin that should be read against the glass/Perspex's printed scale from the middle of the ball bobbin. When gas is flowing through a flow tube the bobbin should continually rotate freely at every flow rate.

Antistatic coatings have been applied during manufacturing in order that the bobbin does not 'stick' in its travel up and down the tube. Manufactures will supply cleaning kits that contain antistatic pipe cleaners, gloves and seals that are to be used during routine servicing. Great care must be taken when dismantling, cleaning and reassembling to ensure there are no leaks, the bobbin moves freely and the printed scaling is perfect. Note: the tubes and bobbins are not interchangeable and must only be replaced with an identical component for the specific machine and gas. On many anesthetic machines the tubes are 'pin-indexed' in order that they can only be fitted in the correct position.

You may also notice that there are often two flow tubes for a particular gas. The first is scaled for 'low flow' (0–1 L/min) and the second is for higher flows (1–15 L/min). This allows for greater accuracy at lower flows, oxygen and air commonly have two flow tubes.

Anesthesia workstation standards require that whenever oxygen supply pressure is reduced, the delivered oxygen concentration at the common gas outlet (CGO) does not fall below 21%

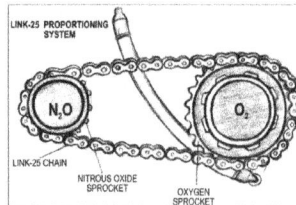

FIGURE 5.2.12   Hypoxia prevention devices (min oxy guard).

One of the 'minimum machine' standards that is incorporated into all anesthetic machines is a requirement to supply a minimum of 20% oxygen (O_2) whenever nitrous oxide (N_2O) is provided to the patient. As has been referred to before, N_2O does not contain oxygen and supplying just N_2O will result very quickly in the patient becoming hypoxic. Most hypoxic guard systems are set to always supply 25% oxygen against any N_2O flow level set. There are several methods used by manufacturers to address this requirement. Two of the most common engineering solutions are as follows: the first is a 'chain-linked' mechanical connection between the N_2O rotameter and a paralleled needle valve at the back of the main O_2 needle valve that will open and supply O_2 when the N_2O rotameter/needle valve is opened. As can be seen in Figure 5.2.12, the circumference of the N_2O cog is about 25% of that of the O_2 cog. This means that a full rotation of 360° of the N_2O rotameter will result in a 90° (25%) rotation of O_2 rotameter. The second solution is a differential regulator (not shown) that allows an increasing bypass flow paralleled around the back of the main O_2 needle valve when N_2O flows, again oxygen is then supplied up through the O_2 flow tube. Today much of this is often achieved by electronic control systems within the machine rather than mechanical controls. Servicing and calibration of both systems is an important task to ensure patient safety.

CGO

This has a standard 15 mm female slip joint fitting with 22 mm coaxial connector. Machine standards dictate that accidental disconnection of the breathing hoses at this point be prevented or made difficult.

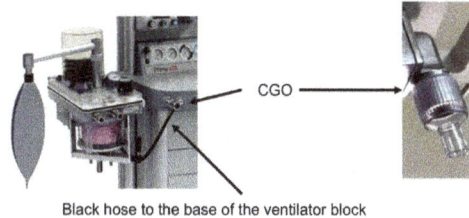

Black hose to the base of the ventilator block

FIGURE 5.2.13   Common gas outlet (CGO).

The port at the front of an anesthetic machine is the feed of FGF to the anesthesia circuit, whether it be a circle system or top circuit. As stated in Figure 5.2.13, there is a 15 mm slip joint fitting that has a 22 mm outlet connector that must be secured to prevent accidental disconnection. On many machines there is a black hose from the CGO to the bottom of the ventilator block.

1. All connections in the scavenging system are of 30 mm diameter, which is distinctly different from the airway accessories (15/22 mm) making misconnections improbable.
2. High Flow, low pressure – 'Air Brake'

FIGURE 5.2.14   Scavenging systems.

As has previously been stated before, the purpose of the scavenging system is to remove any excess anesthetic gasses from not only the patient circuit but also the OR environment. On a conventional anesthesia machine there is usually at the rear of the machine a scavenging system tube that is connected to the anesthesia circuit and also has an open port close to the floor (Figure 5.2.14). Scavenging systems then connect to a wall connector that is often sited along with the medical gasses. The connector for the scavenging system is usually a screw connector that is distinct from the medical gasses connectors. Flow through the tube is indicated by a large disk-shaped bobbin floating upwards in the clear tube. On more modern anesthetic machines, there is often a built-in scavenging system with the hose connector mounted at the rear along with the medical gasses hose connectors and an open port at the bottom of the machine close to the floor to remove any excess anesthetic gasses for the OR environment.

The wall connectors to which the scavenging system connect provide a negative high flow/low pressure draw in order to remove the excess anesthetic gasses either to a roof-top opening that allows the gasses to freely escape to the outside environment or an anesthetic gasses neutralizing system that is placed in the hospital's plant area in the roof space of the building.

6 Anesthesia Patient Gas Monitoring

To understand the principals of measuring oxygen, carbon dioxide and anesthetic gasses, it is important to first understand the basic structure used to draw a sample of gas from the patient ventilator circuit. In almost all cases, the patient is unconscious and is undergoing surgery and thus connected to an anesthetic circuit and mechanical ventilator by a small slip Luer connector (plastic twist connector) on the patient's 'Y' piece. From this, a sample tube is connected that feeds a low flow of approximately 200 mL/min to the patient monitor. Within the monitor are the various electronic modules that undertake the identification and concentration measurements.

Not only is the gas system within the patient monitor measuring patient gasses, but just as importantly, it is measuring the anesthetic machine's output gasses. The gas waveforms on the monitor also reflect the inspiration and expiration phases of breathing whether they are the patient's own spontaneous breaths or mechanically controlled breaths driven by the ventilator.

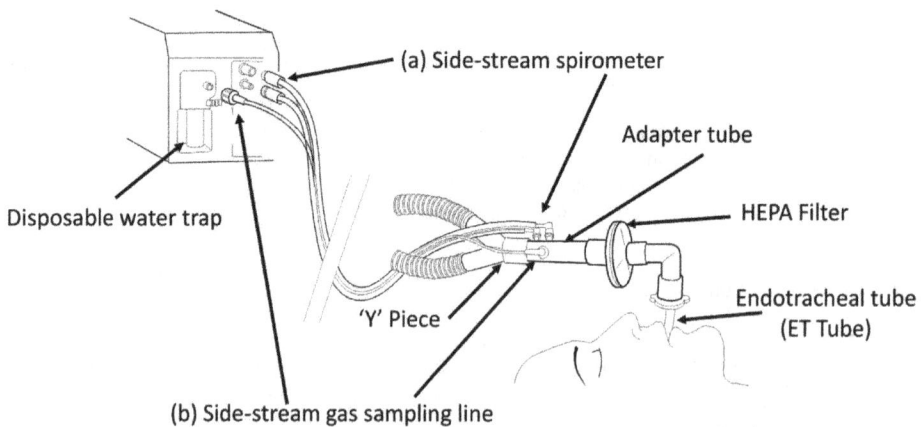

FIGURE 6.1   Side-steam gas sampling and side-stream spirometry.

In Figure 6.1, we see the standard connections between a patient monitor gas measurement module and the anesthetic gas delivery system connected to the patient's endotracheal tube (ET tube). This tube is inserted into the patient's mouth and upper airways in order that they can be mechanically ventilated. Connected to the ET tube is a heat exchange particulate absorber (HEPA) filter. This has two functions: one is to protect the patient's airway from air-born particles such as dust and the other is retaining heat and moisture within the patient's airways.

In Figure 6.1 a dual hose is connected for side-stream spirometer (the measurement of air flows for both inspiratory and expiratory phases), this is connected via an adapter tube that has a small restrictor in the middle. Measuring the pressures on either side of this small restrictor during both this inspiratory and expiratory phase of the breath and allows the patient monitor to calculate the flows, volumes and pressures of the gasses delivered to the patient.

A single tube is also attached to the adaptor and draws a small flow of ~200 mmL/min into the patient monitor gas module. Within the gas module, this gas is then passed through the water-trap

DOI: 10.1201/9781003609414-6

in order to remove any moisture. This gas sample line is routinely changed between patients in order to reduce cross-contamination.

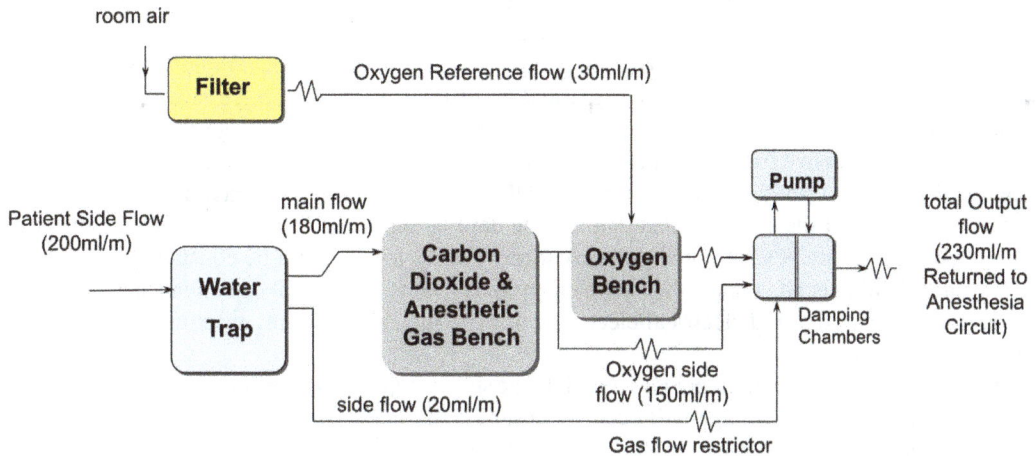

FIGURE 6.2   Typical patient monitor gas measurement flow circuit.

It is important to understand the basic structure used to draw a patient sample of gas from the patient ventilator circuit. One of the first things you may notice when you look at Figure 6.2 is the word 'bench'. This is an historical term and commonly used many years ago when gas measurement devices were literally as large as a table bench. Long before patient gas measurement was available, large medical gas and industrial gas producers would use gas bench analyzers with high flows in order to monitor the concentration of gasses they were producing. With the advent of modern electronics and particularly highly sensitive pressure transducers, medical companies were able to drastically reduce the size of gas benches to the point that they could comfortably fit within a patient monitor.

At the very heart of the system is a pump that draws a gas flow from both the patient's side flow system and room air within the monitor (reference flow). Starting at the left-hand side of Figure 6.2, patient side flow gas is drawn into the water trap and passes through a hydrophobic filter within the trap. This ensures that almost all of the water vapor from the patient gas sample collects in the plastic bottle within the trap. Within the water trap also there is an output port that supplies a side flow that bypasses the two gas measurement benches. This port is set below the main flow and will allow gas that may still be carrying a small amount of water vapor to be drawn off and avoid the gas measurement benches. It is vitally important that the gasses that are allowed to enter the measurement benches are dry as the internal tubing is extremely narrow and prone to being blocked.

Within Figure 6.2 there are four gas flow restrictors that control the flow. The upper output port within the water trap allows the main flow (180 mL/min) to pass to the carbon dioxide and anesthetic bench and then on to the oxygen bench. From the output of the anesthetic bench, the main flow is divided again with only 30 mL/min passing into the oxygen bench, while the oxygen side flow of 150 mL/min is routed around the oxygen bench. Within the monitor there is also an open port to room air intake (oxygen reference flow 30 mL/min). This is used in the determination of the oxygen concentration.

The three flows enter the first dampening chamber that helps smooth the flow. The total flow passes through the pump to the second output dampening chamber, again smoothing the flow. From the gas output port on the monitor, the gas is then returned by a gas sample return tube back to the patient/ventilator circuit.

WHAT IS OXYGEN?

When asked what oxygen is most people would correctly say that it is a gas. It makes up almost 21% of the air around us and is vital for life. Within medicine, it can also be viewed slightly differently. Not only is it vital for life but also as a therapeutic drug that is prescribed and monitored during its administration. Too little oxygen is dangerous (hypoxia) and can rapidly lead to confusion, brain damage, difficulties breathing, rapid heart rates and death. Excessive oxygen is also dangerous. Reduced heart rate (bradycardia), headaches, nausea and vomiting are just a few of the side effects of excessive oxygen (hyperoxia). In neonates, hyperoxia is particularly dangerous as it can rapidly lead to sight damage to the eyes (ROP, retinopathy of prematurity, detachment of the retina at the back of the eye) and also damage the developing lungs. Whenever supplemental oxygen is given, the patient's saturated pulse oximetry (SpO_2) level should be constantly monitored. Great care should also be taken when supplemental oxygen is in use as it can enhance combustion from sparks, cigarettes and electrical/electronic devices such as diathermy machines that are used in the operating room.

Lack of sufficient oxygen (hypoxia) is possibly the single major cause of clinical incidents. Without sufficient oxygen, brain damage will occur in around 4 minutes and death around 6–7 minutes. It therefore follows that fast detection of a lack of oxygen supply to a patient is very important given the very short time period before injury starts to occurs.

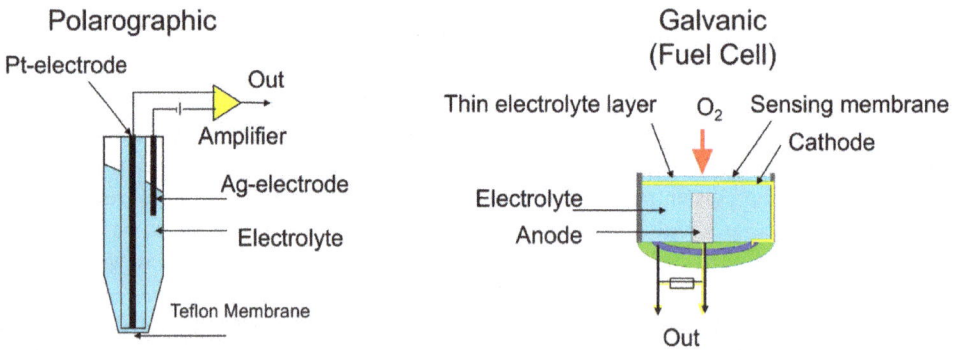

FIGURE 6.3   Different types of chemical O_2 sensors.

The two oxygen measurement devices in Figure 6.3 have for decades been the principal way that oxygen has been measured in medical devices. Without going into the detailed function of these devices, it is sufficient to say that they could almost be considered batteries, whose output current/voltage is dependent on the partial pressure (concentration percentage) of oxygen absorbed through the membranes. A typical output voltage for a galvanic fuel cell is between 7 mV and 28 mV for room air at 21% oxygen concentration. When 100% is applied, the output voltage will increase substantially. Likewise, the output voltage from a polarographic sensor is proportional to the amount of oxygen absorbed through the Teflon membrane.

Both of these types of sensors have drawbacks. First, like batteries, they have a finite life span and have to be replaced and recalibrated regularly every few months. Second, they are reliant on a slow chemical reaction occurring therefore the value that is displayed can only be an average and not an instantaneous real-time value. This has some value, but it is not able to show the distinction of concentration-level oxygen that a patient breathes in against the concentration level of oxygen they breathe out (fractionally inspired, FI and fractionally expired, Fe). A third drawback is that these sensors have to be vacuum packed and have a limited shelf life. They are also relatively expensive, and it is also difficult to manage the routine replacement program needed.

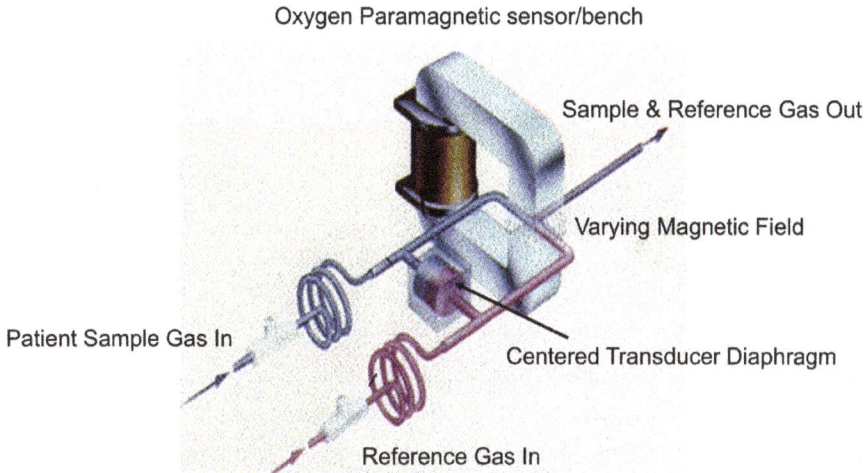

FIGURE 6.4   Paramagnetic oxygen sensor

Oxygen is one of the few gasses that will react to a varying magnetic field. This is due to an oxygen molecule having two unpaired electrons. The result of this is that when oxygen is under the influence of a varying (AC) magnetic field, the oxygen molecules react by changing shape. This change in shape results in a density change of the oxygen molecules for each phase of the magnetic field variation. This density change is sufficient enough to be sensed by a highly sensitive transducer. In Figure 6.4, the diaphragm of this transducer sits between and separates the reference flow of room air and the patient sample gas flow.

If we look at Figure 6.4 what we can see is the following:

1. Both the patient sample gas and reference gas are drawn through the oxygen paramagnetic sensor/bench by means of the pump shown in Figure 6.4, 'Typical Patient Monitor Gas Measurement Flow Circuit'.
2. The reference flow which is room air (21% oxygen) is drawn in within the patient monitor and is used as a reference against which the patient sample gas is compared in terms of partial pressure against the patient sample gas.
3. The two coil restrictors limit the flows in both flows to approximately 30 mL/min. It is vital that these two flows are equal.
4. As the two flows meet and merge within the influence of the varying magnetic field, both produce a varying back-pressure along the internal tubing and on to the centered transducer diaphragm. If it should be that the patient sample gas and the reference gas are both at room air (21% oxygen), then the transducer diaphragm will remain centered.
5. However, should the patient sample gas have a higher or lower concentration of oxygen, the diaphragm position within the transducer will shift to reflect which side of the two flows has the higher concentration of oxygen. This minute shift is sensed by coil pickups within the transducer and the signal passes to the monitor's microprocessor and display.

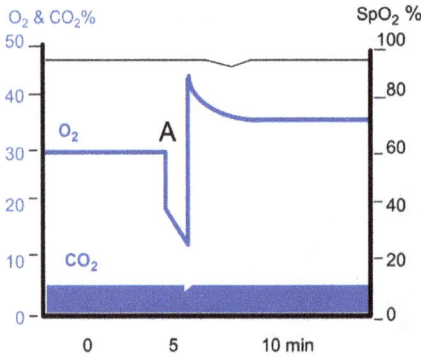

Oxygen instead of nitrous oxide was unintentionally turned off at A. Inspiratory oxygen levels in the circle system rapidly decreased below 18%, which is the fixed alarm level in the monitor.

Note that the alveolar concentration decreased more slowly and never reached a level leading to changes in the capillary blood saturation. A "puff" of emergency oxygen and restoration of oxygen flow immediately restored safe gas levels. CO_2 = carbon dioxide;

SpO_2 = pulse oximeter oxygen saturation

FIGURE 6.5   Circuit hypoxia—a clinical error.

Shown in Figure 6.5 is the display on a patient monitor during a routine surgical operation. On the left side is the scale showing concentrations of both oxygen (O_2) and carbon dioxide (CO_2). The oxygen measurement is showing the concentration of oxygen being delivered to the patient from the anesthetic machine, approximately 30% oxygen. The carbon dioxide reading is a trended level of CO_2 produced by the patient, approximately 4% ~ 5%. These two readings have been sampled as a side flow back from the 'Y' piece close to the patient's mouth and feed to the patent monitor's gas senses/benches. The right-hand scale shows the saturated pulse oximeters (SpO_2) reading of approximately 96%.

The anesthetist, without looking closely at the gas controls on the anesthetic machine, inadvertently turns off the oxygen supply when meaning to turn the nitrous oxide anesthetic gas down. This results in a serious drop in fresh oxygen, below 20% to the patient. However, as can be seen in Figure 6.5 there is only a small insignificant drop in both the SpO_2 and CO_2 readings that fail to alarm. It is only the fast detection of the oxygen sensor within the patient monitor that immediately alarms and makes the anesthetist aware of the issue that is corrected by supplying a puff of 100% oxygen and turning the oxygen supply control of the anesthetic machine to slightly above the original setting of 30%–35%.

CARBON DIOXIDE MONITORING

One of the ways to understand CO_2 monitoring and its importance as a clinical parameter is to think of CO_2 as the output gas from the human body, just as you would think of the exhaust gasses from a combustion engine in a car. In order to obtain the correct reading of the car's exhaust gasses, the air and fuel must be of the correct amounts, the timing of the compression in the cylinders and ignition of the spark plugs must be correctly synchronized. CO_2 that is produced by the human body also requires that systems and functions 'up-stream' must also be functioning correctly in order to produce the correct level of CO_2. These systems and function are ventilation of the lungs, heart and circulation of blood around the body, oxygen absorption by organs and tissues, metabolism within the organs and tissues that as a byproduct produces CO_2. This is then returned via the cardiac circulation to the lungs to be expelled. CO_2 is also one of the minimum monitoring standards set by the American Association of Anesthesiologists and Association of Anesthetists (A of A) and Great Britain and Ireland that must be adhered to when operating on patients.

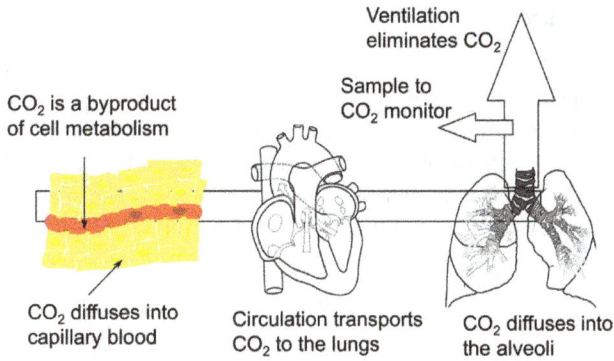

FIGURE 6.6 CO_2 production, transport and elimination.

Starting from the left side of Figure 6.6, oxygen that has been absorbed by organs and tissues is consumed by the process of metabolism. This is the process of burning fats, carbohydrates and proteins in chemical reactions to produce energy, heat, growth, cell replacement, fight infections and other necessary elements to support life. The byproduct is that metabolism also produces CO_2. This is absorbed into the cardiac circulation that returns it back to the right side of the heart and from there to the lungs.

The blood returning to the lungs is low in oxygen but high in CO_2. The air sacs (alveoli) within the lungs are filled with air which is rich in oxygen (21%) and almost zero CO_2. The flow of returning blood is passed around the alveoli air sacs. The walls of the alveoli are two or three cells thickness and have the property of allowing CO_2 gas molecules to pass from CO_2-rich returning blood flow, through the walls of the alveola and into the air within the air sac. In the opposite direction, oxygen that is high within the air in the alveola is able to pass in the opposite direction through the walls of the alveoli and into lungs blood circulation and then returns to the left side of the heart. In order to fully understand the process of gas exchange within the lungs, it is advisable that you first study Chapter 4, Respiratory System and Mechanical Ventilation.

...CO_2 Production, Transport and Elimination 2

- $EtCO_2$ is a non-invasive, breath-by-breath indicator of:
  - Metabolism: production of CO_2
  - Circulation: transport of CO_2
  - Ventilation: elimination of CO_2

FIGURE 6.7

The three points in Figure 6.7 enforce the statements in Figure 6.6.

When discussing the concentration of CO_2 expelled in breathing, the term 'end-tidal' (Et) is almost always used to be placed before the value of CO_2 percentage (i.e., $EtCO_2$ 4.6%) When thinking about the process of breathing, it is possible to think of the air entering and leaving as a tidal motion of gas, in the inspiratory phase and out on the expiratory phase. The very last gas on the expiratory phase will have come from the deepest part of the lungs and will contain the highest concentration of CO_2. Hence, the use of the term $EtCO_2$ refers to measurement of CO_2 concentration at the end of the expiratory phase.

TRANSCUTANEOUS CO_2 MONITORING

You may also have come across the term $TcpCO_2$ (transcutaneous, measured through the skin). This is a separate method of measuring the level of CO_2 in the body by using a heated sensor attached to the skin that has CO_2 lying just beneath the skin to defuse up through the skin and vary the impedance between the anode and cathode of the sensor. This method of CO_2 measurement is often found in neonatal care units and respirator care units within a hospital. The sensors used are extremely expensive and very delicate, requiring great care when handling.

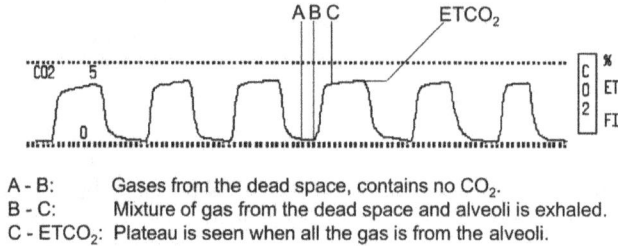

A - B: Gases from the dead space, contains no CO_2.
B - C: Mixture of gas from the dead space and alveoli is exhaled.
C - $ETCO_2$: Plateau is seen when all the gas is from the alveoli.

FIGURE 6.8   The origin of the CO_2 waveform capnography.

Capnography refers to the monitoring of the production of CO_2 when breathing. The real time capnographic waveform in Figure 6.8 shows the usual waveform of concentration of CO_2 shown on a patient's monitor in the operating room.

A–B: Gas from the dead space, containing *no* CO_2. This is the first gas to be expelled during the expiratory phase and has resided in the patient's upper airways such as the mouth, trachea and the left and right bronchus. As these air way passages are not involved in the gas exchanges, they are therefore known as 'dead space'.

B–C: Mixture of gas from the dead space and alveoli is exhaled. It is here that we have a blend of air that is part dead space gas and also some gas that had been involved in gas exchange in the upper parts of the lungs.

C–EtCO₂ (End Tidal Carbon Dioxide): Plateau is seen when all the gas is from the alveoli. This is the point at which air from the deepest part of the lungs that contains the highest concentration of CO_2 is expelled. The measurement that is displayed on the patient monitor is the value measured the instant before the start of the inspiratory phase.

So far, we have only discussed CO_2 values being measured and displayed in percentages (%), this is the most common unit of measurement used in the United Kingdom. Countries around the world have adopted other units of measurement. These are mmHg (millimeters of mercury) and kPa (Kilopascals).

As an example, CO_2 value 5% = 38 mmHg = 5.06 kPa.

Capnography Definitions

Normocapnia	ETCO₂	4.5 - 6.0 vol % 34 mmHg - 45 mmHg
Hypocapnia	ETCO₂	less than 4.5 vol % 34 mmHg
Hypercapnia	ETCO₂	more than 6.0 vol % 45 mmHg

FIGURE 6.9

The correct level of CO_2 circulating in the blood stream has an important impact on the functioning of several organs such as the brain, heart and kidneys. The levels shown in Figure 6.9 are the generally accepted levels of $EtCO_2$ values. One other very important point is when the CO_2 level rises from the normocapnic level, there is an increase in the acidic (acidotic) levels in the blood stream and conversely if the level of CO_2 drops, the blood will become alkaline (alkalotic). It is the impact of both the acidotic and alkalotic reading in the blood stream that has an adverse impact on organs throughout the body.

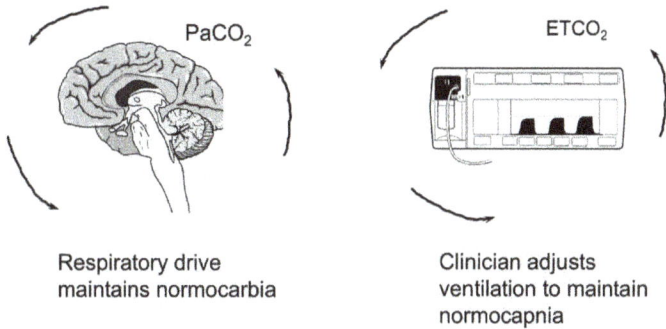

PaCO₂

ETCO₂

Respiratory drive
maintains normocarbia

Clinician adjusts
ventilation to maintain
normocapnia

FIGURE 6.10   Using ETCO₂ to control ventilation.

RESPIRATORY DRIVE

During normal breathing the level of CO_2 in the blood stream play an important part in the regulation of respiration rate and heart rate. At the rear of the brain, close to the brain stem is an area called the medulla oblongata (medulla). This area has a CO_2 monitor built that detects the level of CO_2 in the blood stream and provides control signals to the heart and lungs. If we consider the situation where a person undertakes vigorous exercise and dramatically increases metabolic activity, then, as expected, the consumption of oxygen increases that then leads to a sharp rise in CO_2 in the blood stream. On detecting this sudden rise in CO_2, along with other signals from heart, the medulla then signals to the heart to increase its heart rate and stroke volume (the amount of blood flow per beat) and also signals the lungs to increase its rate and tidal volume (the volume of air per breath). This is something we have all experienced.

During a general anesthesia procedure where the patient's own respiratory drive is overcome by the effect of the anesthetic agents it then becomes the responsibly of the anesthetist (anesthesiologist) to provide control of the level of CO_2 by means of the patient monitor capnograph readings, ventilator settings for fresh gas volumes, flows and pressures (Figure 6.10). The anesthetist may also use drugs that stimulate and control the heart in order to achieve normocapnia and maintain a steady state during the surgical procedure.

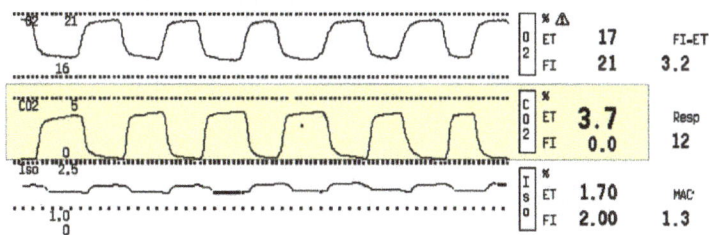

ET	17	FI-ET	
FI	21	3.2	
ET	**3.7**	Resp	
FI	0.0	12	
ET	1.70	MAC	
FI	2.00	1.3	

Alveolar minute ventilation is usually adjusted to achieve normocapnia,
where ETCO₂ is in the range 4.8 - 5.7 % (36 - 43 mmHg).

FIGURE 6.11   Normocapnia and normoventilaltion.

Figure 6.11 shows what might be displayed on a patient monitor in the operating room during surgery—the three waveforms of oxygen (oxygraph), carbon dioxide CO_2 (capnography) and the anesthetic agent graph. Here is a brief explanation of each of the three wave forms:

Oxygen: As the patient is breathing air, the scale for the graph is set with a maximum of 21% to a minimum of 15%. The concentration of oxygen in the air being 21% FI (fractionally inspired) is shown below as a value beneath Et (end-tidal) value of 17%. A calculated value of FI–Et shows an uptake of oxygen by the patient.

CO_2: The waveform for CO_2 appears to be a complete inverse of the oxygen wave form. A scale of 5%–7% maximum and a minimum of 0% are the norms. The figures to the right of the waveform show CO_2 of FI 0.0% and Et 3.7%. This is therefore showing that at this point in the surgical procedure the patient is slightly hypercapnic. Also being shown next to the CO_2 values is the respiratory rate of 12 breaths/min.

AA (anesthetic agent), which in this example is the AA isoflurane (Iso) is showing that at this point in the surgical procedure, the anesthesia machine is delivering a FI 2% Iso and the patient is consuming 0.3%, so the value of Et Iso is given as 1.7%.

You will also notice that a value of 1.3 is given for minimum alveolar concentration (MAC). This is a crude calculated scale used to show the potency of all the commonly used anesthetic gasses and vapors used for anesthesia. A simple explanation is as follows: during early trials of each anesthetic vapor or gas, 100 people are administered a particular anesthetic agent at a given concentration of 1.4%, and of these, 50 people do not show any response to a surgical incision and so for that particular anesthetic agent it is said that for a concentration 1.4% it has an MAC value of 1. This is particularly useful when mixed agents such as Nitrous oxide (N_2O) and isoflurane (Iso) are being administered during surgical procedures as the combined effects of the two anesthetic agents can be managed. MAC is today not the only method of monitoring the effect of the anesthetic agents on the patients.

Sudden Disappearance of CO_2 Waveform

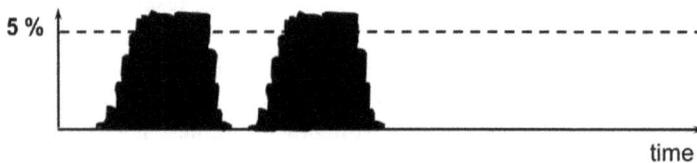

If the CO_2 waveform is normal but then drops to zero from one breath to the next, the most common cause is a ventilator disconnect.

A complete airway obstruction, for example caused by a fully kinked ET-tube, is another possibility.

FIGURE 6.12   Sudden disappearance of CO_2 waveform.

Figures 6.12–6.16 are examples of how a CO_2 (capnographic) waveform can be used to interpret clinical situations that may occur during surgery. One of the most useful features of the CO_2 waveform is that it is almost instantaneously changes when problems occur.

Esophageal intubation

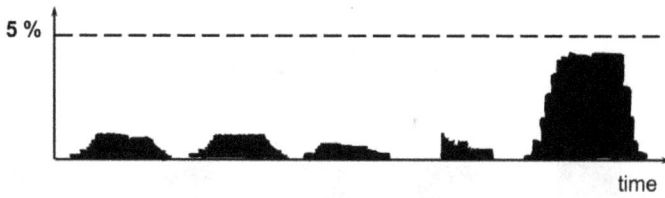

Some CO_2 may be detected due to gas entering the stomach during manual ventilation. After removal of the endotracheal tube and successful intubation, high $ETCO_2$ values are recorded due to CO_2 accumulation during apnea.

FIGURE 6.13

In Figure 6.13 you can see how the CO_2 waveform is used to determine that the ET tube has been inserted incorrectly at first on the wrong side of the epiglottis toward the stomach. The ET tube was then withdrawn slightly and using a laryngoscope repositioned into the trachea. As can be seen, the CO_2 waveform returns to normal.

Lack of Relaxation

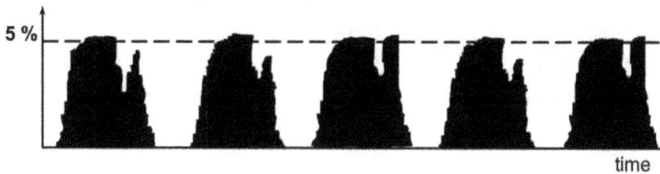

Insufficient muscular relaxation and inadequate depth of anesthesia are allowing the patient to "fight against the ventilator", which can be seen as clefts on the capnogram plateau.

FIGURE 6.14

Yet again the CO_2 waveform proves useful, this time in showing that the patient is not sedated sufficiently during surgery (Figure 6.1.14).

Partially Obstructed Airway

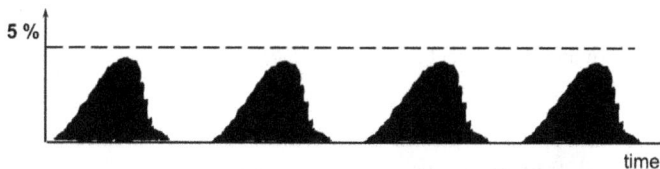

A deformed capnogram, with a slowly rising leading edge, may indicate a partial obstruction of the airway. Possible causes of obstruction are: bronchial asthma, bronchospasm, mucus in the airways or a kinked endotracheal tube.

FIGURE 6.15

In observing such a waveform as the one in Figure 6.15, the anesthetist is most likely to first apply suction to the airway in order to remove any mucus that may have built up.

Rebreathing CO_2

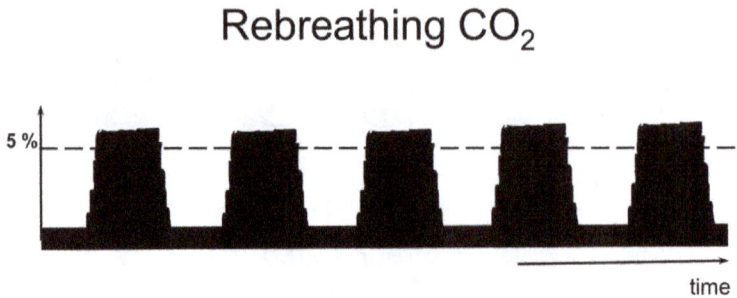

Failure of the capnogram to return to the baseline indicates rebreathing of the exhaled CO_2. It may be due to an exhausted CO_2-absorber in a circle system (soda lime cannister which is part of the anesthesia machine cycle system), or insufficient fresh gas flow rate from the anesthetic machine.

FIGURE 6.16

As you are now aware, delivery of sufficient air flow to the patient during anesthesia ensures that $EtCO_2$ levels remain within normal limits (normocapnia). If the fresh gas flow is insufficient, then there is a strong likelihood of the patient being unable to expel the required amount of CO_2 and this leads to rebreathing of the CO_2 (Figure 6.16). Rebreathing will then lead to hypercapnia. There can be a number of reasons for CO_2 remaining and being inspired during the inspiratory phase such as an exhausted CO_2 soda lime cannister in a circle system. This will be covered in the section within this book on anesthesia.

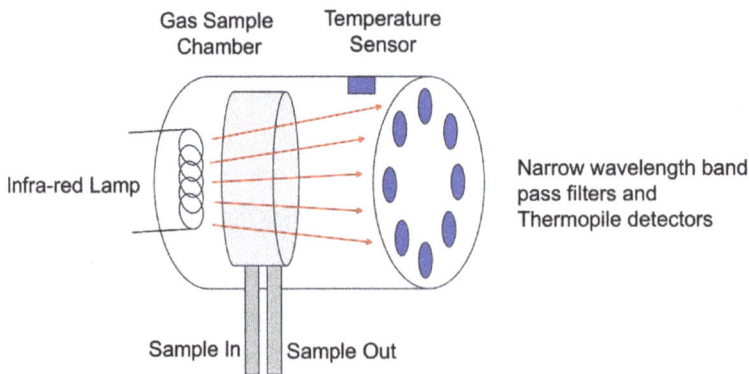

FIGURE 6.17   The modern anesthetic agent and CO_2 gas measurement bench.

Although AAs are not carbon-based gasses/vapors, they do however have the property to absorb infra-red light at spectral frequencies very close to those of CO_2. This property has enabled the design of the modern AA bench which incorporates the measurement of both AA and CO_2. The Figure 6.17 shows a simplified diagram of a modern AA and CO_2 gas measurement bench. The infra-red light is shone through the gas measurement chamber. It then passes through individual narrow wavelength band pass filters before striking the face of the individual thermopile detectors.

Each thermopile produces a signal that reflects the level of infra-red absorption at a particular infra-red wavelength. Each of the anesthetic gasses/vapors and CO_2 produces a distinct pattern of signal-level outputs from the thermopile detectors, the patient monitor software is then able to identify an individual anesthetic gasses/vapors and CO_2. It is also able to determine from the signal levels the concentration of each. One of the major advantages of the modern AA/CO_2 bench over the earlier benches is that it does not contain any moving parts.

7 Assessment and Care of the Newborn

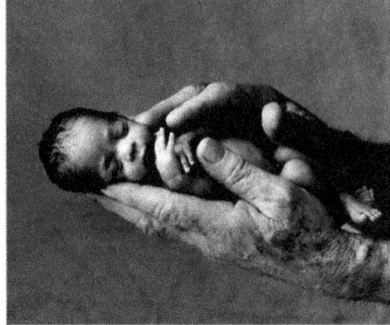

Of all the clinical disciplines that a clinical engineer can find themselves working in during their career, it is difficult to imagine one more rewarding than working to support new life. Caring for such small and vulnerable human beings requires a high level of professionalism and commitment by all working in the neonatal field. This can be emotional and personally challenging for all staff, from the most junior nurses to the most senior doctors and consultants. Not only are the care demands of the babies extremely high, but there is also required a high degree of commitment needed to support the baby's parents and families during what are often the most stressful of times. The day a baby leaves hospital into the care of their parents, having been very unwell during their early hours and days, is a day that is often quietly celebrated by all involved in the baby's care. That can extend to the clinical engineers working in the background within the neonatal department who have ensured the safety and availability of medical equipment that plays a pivotal role in the caring and therapy of these little human beings.

One other point that should be considered is that of the baby's size compared with the other two of groups of patients, pediatrics and adults. A newborn's weight is typically just 5% of that of an adult and as such much of the equipment used to care for them is scaled down in size in order to account for this. Tubes, electrodes, cots and incubators are just some of the medical equipment specifically designed for the care of neonates. This does mean that medical staff working with babies are required to be in very close proximity to the baby; this is typically little more than half a meter away. Along with all the required equipment such as ventilators, humidifiers and patient monitoring this produces often very cramped conditions in which medical staff have to work.

As a clinical engineer working in the neonatal area, you should also take extra care in your approach in order to ensure that you do not inadvertently cause concerns with parents and staff. This can be difficult when you need to gain access to equipment as parents very often do not understand your role and sometimes feel their privacy is compromised by a stranger working in the department. Speaking quietly, smiling and a general positive demeanor all help in reducing parents' concerns.

DOI: 10.1201/9781003609414-7

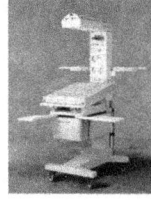

New born Assessment

- Keep them WARM!
- Respiratory rate
- Baby's ability to keep airways clear
- Examination of infant for abnormalities
- Determination of temperature, heart rate and often blood pressure
- Analysis of new born 'Maturity'
  - Oedema, Skin texture, skin colour, ear formation, genitals etc
  - Apgar Scoring

FIGURE 7.1

Let's start by considering a comparison of the environmental changes in the few short hours before a baby is born and the minutes that follow their birth. Within the womb they are in a dark, quiet, warm (~37°C), tightly enclosed environment surrounded by wet amniotic fluid. In the minutes following their birth they find themselves in a bright, noisy, cooler (~30°C) external temperature without as much physical bodily support, and, of course, a dry atmosphere. They now for the first time hopefully draw their first breaths and begin to live independently of their mother.

The first few minutes of any new born baby's life is always a time of some anxiety for both parents and medical staff. Ensuring the well-being of baby and mother following the birth requires experience, knowledge and skill from the attending medical staff to ensure the safety of both baby and mother (Figure 7.1). For baby, it starts with the most important step, that is, to dry the baby and keep baby warm. It might be surprising to know that keeping a baby warm is the number one step before even accessing their breathing. Next, a detailed visual examination of the baby is undertaken to check for any abnormalities such as talipes (club foot), cleft lip and/or cleft palate and scoliosis (deformity of the spine). If there should be some concerns over the baby's condition, measurements of heart rate, temperature and possibly blood pressure are taken. Should the baby be considered either premature or low birth weight, the clinicians will often undertake to score the baby's maturity by using such an assessment of skin texture, skin color, ear formation and genitals. These checks are often undertaken as the baby lies on a mattress under a radiant warmer (Figure 7.2).

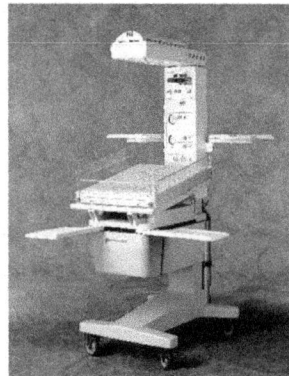

- A purpose build cot containing the following:
- Radiant heater with Air/baby mode
- Supplemental oxygen supply from wall supply or cylinder
- Low negative pressure suction
- Tiltable mattress
- Apgar clock either mechanical or digital electronic.
- Storage for resuscitation equipment and blankets

FIGURE 7.2   Radiant heat warmer/intensive care cot (Resuscitaire).

Today's modern radiant warmers allow for the control of the radiant heater by either setting the heater power (air mode) or by attaching a digital thermometer probe to the baby's skin surface that will adjust the heater output to control the baby's skin surface temperature to a 'set point' set by the clinician. The radiant heater is the single most important component of the cot and should be turned on for some time prior to baby being placed on the mattress. Clinicians should always be aware when leaning over to work with a baby that they do not obscure the radiant heat from the baby. There is also a potential risk that if the clinician is wearing any form of head covering themselves, this potentially can catch fire should it inadvertently touch the elements of the radiant heater. To the front of the radiant heater head there is usually a light that will illuminate the baby lying in the cot in order to allow for examination of the baby. It only emits very low heat and so does not contribute to warming baby.

Supplemental oxygen is also available for babies who have poor respiratory effort and/or low oxygenation saturation (SpO_2). This oxygen supply should where possible be from the wall supply via an oxygen hose to the resuscitaire or from the attached oxygen cylinders at the rear of the resuscitaire. When the resuscitaire is not in use, then it is always advisable that the cylinders are checked to ensure they are full by checking the oxygen cylinder contents gauge on the front, and then using the cylinder key to ensure that the cylinders are turned off, so they are available when next needed. Low-pressure suction is also provided in order to clear baby's airways of any residual fluid when born (mucus). This low-pressure negative suction is derived from a Venturi negative pressure system within the resuscitaire using either oxygen or air. The maximum pressure of the suction system is very low and great care is required when using suction in order to not damage any of the baby's airways. In some neonatal departments, this low-pressure suction is provided by a wall gas outlet negative pressure controller and collection jar. To aid with removal of mucus from baby's airways, the mattress has the option to tilt with the baby's head slightly below their feet. This is known as the Trendelenburg position and aids in removing mucus and providing oxygen.

A timer is also integrated into the resuscitaire and is used during scoring the baby against the Apgar scoring system (Figure 7.3). This will provide an audible alert at 1 minute and every 5 minutes following. The next figure provides more details on the Apgar scoring system. Storage of all the required equipment that maybe required for a baby's resuscitation is stored in the draws below the mattress. These will include a stethoscope, towels, suction tubing and oxygen face masks along with many other items. As with the cylinders, it is important that the stock levels of these items are regularly checked and any deficiencies replaced.

Apgar Scoring
at 1 and 5 minutes

SCORE	0	1	2
Heart Rate	Absent	< 100	> 100
Respiration	Absent	Irregular or gasping	Regular sustained
Tone	Limp, flaccid	Some flexion of extremities	Good flexion and tone
Reflex (response to suction)	No response	Weak grimace	Cough, sneeze or cry
Colour	Blue or pale	Lips pink blue extremities	Pink all over

FIGURE 7.3 Apgar scoring at 1 and 5 minutes.

In cases where baby appears to need immediate medical support, an assessment known as Apgar scoring is quickly undertaken within the first minute of baby's birth. This scoring system was first devised by Dr Virginia Apgar in 1953 in the United States. It is now a worldwide adopted method of evaluating a baby's condition within the first few minutes following the baby's birth. The scoring system takes place at set time periods of the first minute and then every 5 minutes until the baby has recovered or the resuscitation of the baby is stopped often around 20 minutes from birth.

Five parameters are assessed at each time interval with each parameter being given a value of 0, 1 or 2. The total score is tallied each time and hopefully an increase in total score is seen as the 5-minute intervals increase. A score of 7+ is usually acceptable and baby's often without further intervention continue to improve. Scores of less than 7 often require clinical support such as supplemental oxygen, suction and additional warmth, all of which can be provided via the radiant warmer (resuscitaire).

During the first few weeks of life, babies should always have a heart rate in excess of 100 b/pmin. As babies develop over the coming months and years, this rate will begin to reduce. The heart rate can be determined by one of two methods, either by the use of a stethoscope listening to the heart or a saturation monitor (SpO_2). Without the detection of a pulse baby is very much at risk of dying. Certain steps such as CPR, with the clinician using just a few fingers to compress the chest, can be attempted along with the administration of drugs that will hopefully stimulate the heart in order to obtain a pulse and restore circulation.

Effective respiration is vital in ensuring the oxygenation of organs such as the lungs, brain, liver and kidneys until they begin to function normally. One advantage that babies have is that their hemoglobin within the blood has the capability of carrying additional oxygen molecules compared to children and adults. This additional store of oxygen allows for a slightly longer time of hypoxia (lack of oxygen) before serious negative impacts occur. Failure to breathe can be treated by both physical stimulus, use of a ventilator and drugs. Babies often suffer from intermittent apnea (cessation of breathing) events during the first few minutes and hours following birth and should always be carefully monitored.

Muscle tone is accessed by observing the baby's physical movements and posture, particularly when stimulated by handling. Hands and leg movement are just two indicators that are looked for and a clinician might open a baby's palm in order to see if the fingers return to a closed position. The most concerning observation is that baby is 'floppy' and completely lacking in responsive movement.

Reflex is somewhat different to muscle tone as it is the baby's autonomic (involuntary) response to stimulation such as suction or handling. Changes in facial expressions such as grimace, sneezing, cough and crying all provide a means of determining reflex from stimulation.

Immediately in the seconds following a baby's birth, an assessment of baby's skin color can be determined. A well-oxygenated baby will have a slightly pink color to the torso (chest, abdomen, pelvis and back), head and extremities (legs and arms). If baby shows signs of 'cyanosis', a bluish or pale color, this can be due to a number of reasons including a congenital heart defect, such as a hole within the heart that has not sealed. Often this can resolve itself within a few days or it can be treated with a surgical procedure.

Physiology parameter
values for new born

- Temperature: 36.6°C – 37°C (97.7°F – 98.6°F)
- Heart Rate: 110 – 160 beats/minute
- Respiratory rate: 30 – 60 breaths/minute

FIGURE 7.4

The values in Figure 7.4 represent the normal range for the baby's vital signs that would indicate a generally well-baby. This of course does not mean that baby is in perfect health as baby may have other medical problems such as jaundice and Down's syndrome. The medical problems that can affect a new born baby are many and varied and attending clinicians are constantly accessing baby for such problems not just by reference to Apgar scoring but using blood tests and medical observations.

As has already been stated, baby's temperature is a critical measurement and as you can see, it is a very narrow window of normal—36.6°C to 37°C. A regular heart rate of between 110 and 160 b/min is considered normal. A rate of below 100 b/min for a newborn baby is said to be bradycardia and a rate of more than 160 b/min is referred to as tachycardia. Tachycardia often resolves very soon after birth, but can persist due to electrical conduction problems within the hearts ventricles and if baby is in pain.

Observing a new born baby breathing, you may first notice that unlike children and adults there is little movement of the chest but their tummies/bellies (abdomen) muscles are used to assist the diaphragm muscle, this is commonly known as belly breathing. It may also be observed that as the belly expands, the chest may fall slightly in what appears to be a 'seesaw' breathing pattern. As you may know an adult breath rate at rest is ~12–15 b/min, but in a newborn baby this is considerably higher between 30 b/min and 60 b/min.

As you can see, the energy demands for a newborn baby are very high, and in order to maintain this, high metabolic activity requires oxygen, food and warmth. One other important indicator is baby's birth weight. We have always understood that a baby's birth weight is an important indicator and, along with baby's sex, is something that parents want to share with family and friends. It is normal during the first few days of life that a baby's birth weight will drop a little, but as baby begins to thrive, they regain this temporary loss of weight and continue to grow at a rapid rate.

- Premature: gestational age less than 37 weeks
- Mature: gestational age 37 – 42 weeks
- Postmature: gestational age greater than 42 weeks

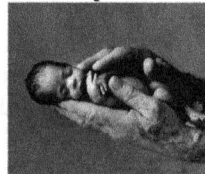

FIGURE 7.5 Assessment for maturity.

For many decades, it has been possible to estimate with some accuracy the date of conception using such information as mothers' last date of menstruation, date of positive pregnancy test and assessment of the earliest ultrasound scan. These factors give a close but not absolute date of conception and are usually given as 'You are about 12 weeks pregnant'. From the date of conception, the 'due date' some 40 weeks later is estimated. This date can be redefined as the pregnancy continues to improve due date accuracy.

A baby can be assessed as premature should baby be born earlier than 37 weeks' gestation (the period of time between conception and birth), mature if born between 37- and 42-weeks' gestation and postmature if born after 42 weeks' gestation (Figure 7.5).

Gestational age

- It is important to understand that size alone is not always a direct indication of gestational age – as child can be small without necessarily being premature

- Three classification help in respect
  - Small for gestational age (SGA)
  - Appropriate for gestational age (AGA) or normal for gestational age (NGA)
  - Large for gestational age (LGA)

FIGURE 7.6

The terms listed in Figure 7.6 are commonly used by clinicians when discussing baby's gestational age. Many low-birth-weight babies are perfectly developed without underlying medical problems. There are however a number of factors that can influence baby's gestational age such as if mother has had a poor diet, smoked or there was placenta dysfunction (the placenta is a temporary organ that connect mother's uterus to the baby within the womb, umbilical cord prolapses). LGA (see Figure 7.7) can occur if mother herself is obese and/or had LGA babies before. There are also some other genetic factors that can contribute to this.

- If TE rises above the value of A, the body temperature rises above the normal range

- If TE falls below the value of B, the body temperature remains within normal range but the baby's heat production increases to compensate for the increased heat losses

- If TE falls below the value of C, the baby can no longer increase his heat production and his body temperature falls below the normal range

- Question – Where is the 'Thermoneutral range'?

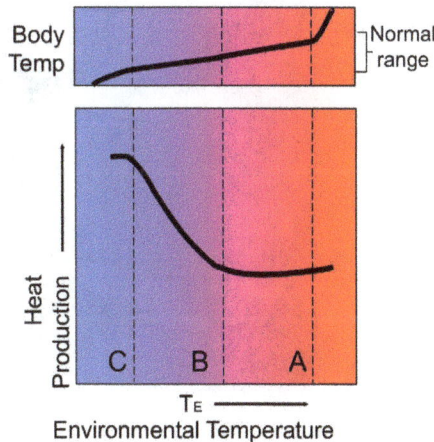

FIGURE 7.7   Body temperature–heat production–environmental temperature.

THERMAL STRESS AND HEAT BALANCE

Careful reading of Figure 7.7 will demonstrate just how important it is to attain a thermal neutral environment, where baby is neither losing heat to the environment nor gaining heat from the environment while producing just enough heat to maintain their own constant temperature close to 37°C. The answer to the question 'Where is the thermoneutral range?' is of course 'B'.

If baby experiences a cooler environment, then energy that would have been used for development and growth is consumed to produce additional heat and can in the long term mean a 'failure to thrive'. Babies are born with a store of 'brown fat' around their necks and along the aorta which provides additional energy when extra heat is needed but this can be quickly consumed in a cold environment. If the environmental temperature is excessively low, then there can be even more serious impacts that can even lead to death within a few short hours. Signs for experiencing a low environmental temperature by a baby are that they attempt to curl up by bringing their legs and arms closer to their bodies in order to reduce their skin exposure to the environment and will often cry vigorously. Their skin color may become pale as their body reduces blood flow to the skins surface (vasoconstrict).

Excessive heat also will have a damaging impact of baby's well-being and can lead to apneas and dehydration. A common indication of a baby suffering heat stress is they adopt a 'sun-bathing position' where they spread their arms and legs to their fullest extent in order to maximize heat loss to their environment. Their skin color will often change to become red as they increase blood flow to their skin surface in order to dissipate more heat (vasodilation). The most common reason for heat stress within the neonatal unit is excessive temperature settings on incubators, radiant warmers and phototherapy units.

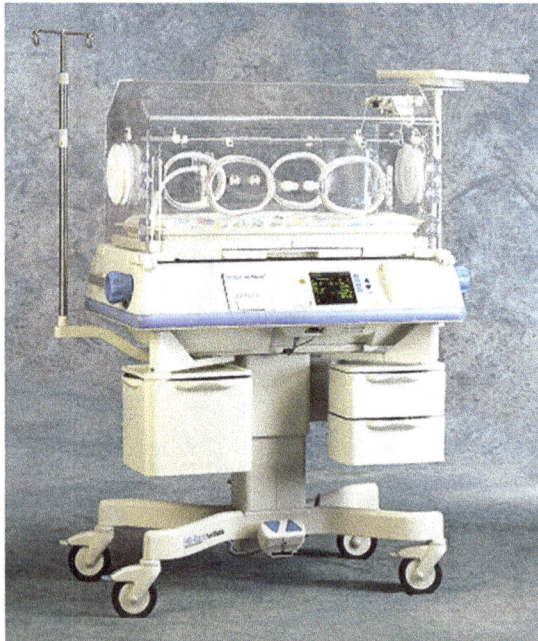

FIGURE 7.8

Baby incubators, often referred to as neonatal incubators, are possibly the single most important device in the neonatal area of neonatology (Figure 7.8). Remaining warm at the appropriate temperature is paramount for all babies in those early days of life. Very accurately controlling the thermal environment when baby is placed in an incubator is something that should never be taken for granted. As clinical engineers, we are often asked to service and repair incubators and when we do, we should always take great care to ensure their correct functioning and accuracy.

Door seals and latches must always be inspected and changed if in any way damaged. The use of 'calibrated' digital thermometers in order to ensure accurate and steady internal incubator temperatures are essential tools. It is not possible to ensure the accuracy of the air temperature solely by relying on the incubator's own temperature displays. As incubators have electronically controlled fans in order to circulate the air within the incubator, it is advisable that air flow meters (anemometers) and sound meters (decibel meters) are used in accordance with the manufacture's instruction.

From experience, I would always undertake a simple check of the sound within the canopy by placing my ear closely to an open canopy door in order to not only check on the base level sound but also to identify any ticking sound that may be generated by the air flow motor. These ticking or other sounds are generally not detectable by a sound meter but would prove to be stressful to the baby. A visible inspection is very important and checks of such things as the caster wheels, Perspex canopies and clarity of displays along with full function checks are recommended and set out in the manufactures service manuals.

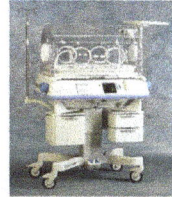

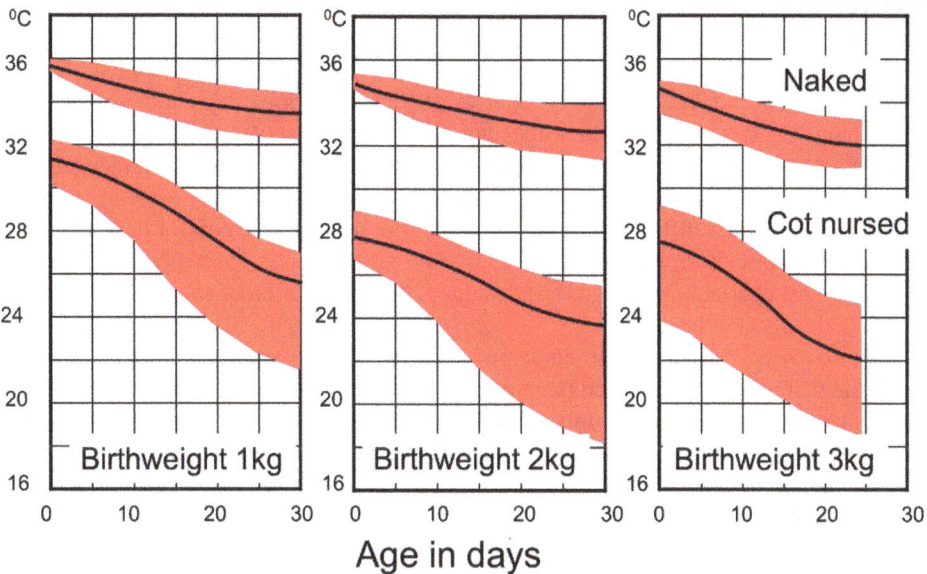

FIGURE 7.9   Correct incubator temperature determined by birth weight.

The question of what should an incubator temperature be set to, sometimes referred to as the 'set-point' is determined by a number of factors such as baby's weight, gestational age, age in days and any underlying medical problems. Figure 7.9 should only be seen as a general guide but shows taking into account these factors and importantly whether baby is clothed (cot nursed) or naked. From these factors, the air temperature within the incubator can be set with precision in increments of just 0.1°C. In Figure 7.10, the black line running through the red areas is what might be thought as the ideal temperature, but clinicians may for various reasons decide to alter the set point within the red areas. As can clearly be seen, with the increase in days from the date of birth, baby's development generally means that they are increasingly able to tolerate the lowering of the incubator's temperature.

Ideally the baby would be clothed where possible as this will allow for a reduction in incubator temperature. The reason for allowing baby to be nursed naked is that it allows clinician visibility of baby in case of fever due to infections, and easier access for medical procedures such as taking blood. Often you may see that a baby is wearing a small cap or bonnet which aids in the retention of the baby's heat while still offering visibility and easy access.

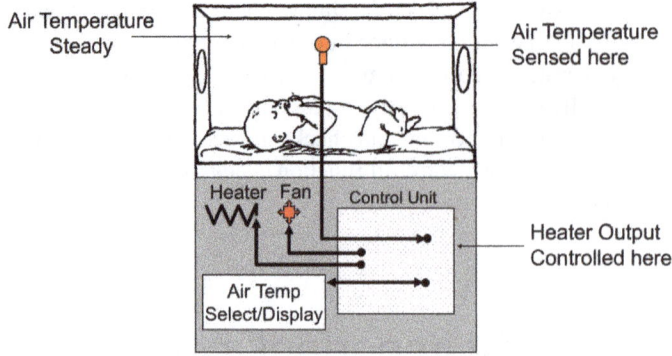

FIGURE 7.10   Air mode incubator.

Please note that different manufacturers adopt different approaches to controlling air temperatures and what follows is only a general guide. Most incubators have the option of choosing between two modes of operation: air mode or baby/skin mode (servocontrol mode, SC). Air mode is the most commonly used mode used as it offers some advantages such as greater air temperature stability when the canopy doors are opened, and does not mask the appearance of a fever in the baby by reducing the air temperature (Figure 7.10). The digital display at the front of the incubator allows the clinician to see what the set point has been set to and the real time measured air temperature. By using the signal obtained from the first air temperature sensor, the control unit will increase/reduce the heater power to track the desired set point temperature. The fan is set to a constant speed and operates continually while the incubator is operating in order to circulate the air from right to left (baby's feet to head).

There is also a second independent temperature sensor, the overheat sensor, mounted within the sensor module that is set to ensure that the air temperature never exceeds 39°C. In some models of incubators there is an option usually only known by senior clinicians to increase the overheat limit to 40°C for babies that are having difficulty maintaining their core temperature. Modern incubators also have the option to monitor and display the baby's skin temperature by using a thermistor probe attached to the baby's skin, often on the abdomen and then using a lightweight cable that plugs into the incubator display at the front of the incubator. The probe is usually attached using a round adhesive disk with a reflective silver coating that reflects away any heat from within the incubator and only senses the skin temperature of the baby below the sticky round disk. By being able to display the baby's skin temperature, the clinicians are able to quickly see if baby is becoming hypothermic or hyperthermic and thus react quickly to the clinical change.

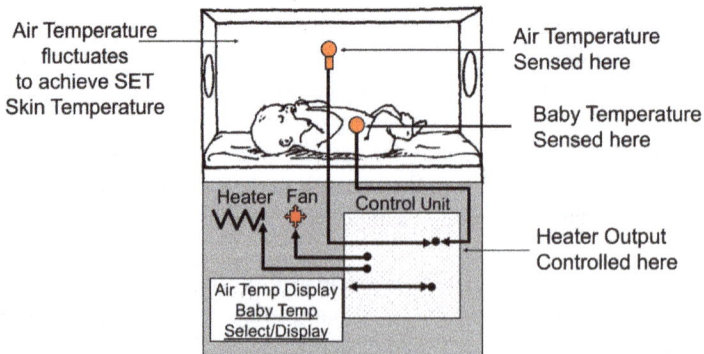

FIGURE 7.11   Baby mode incubator.

Again I have to stress that different manufacturers have slightly different approaches in exactly how baby mode operates. Baby mode, which is often referred to as servo mode, is a mode that uses the baby's own body temperature to control the amount of heat from the heater (Figure 7.11). Primarily used for the smallest and most unwell babies, it offers the ability to gently raise the baby's temperature by ensuring that the air temperature is only about 1°C above that of the skin temperature until the point when the set point for baby's skin temperature is reached. Should the baby's skin temperature begin to exceed that of the set point skin temperature, then the heater power will be reduced allowing the baby to cool by a small amount. The advantage of this mode is that should the baby become pyrexial (raised body temperature due to fever), then baby will be able to lose some of the excessive heat to the air within the incubator and remain comfortable and not heat stressed. The disadvantages are that a baby's fever becomes more difficult to detect and there are great fluctuations in air temperature when canopy doors are opened.

Incubator Heat loss - Convection

- Loss of heat to the surrounding air

- Dependant on the difference between Air and Skin temperature

- Air speed also effects loss

- Clothing the infant can reduce loss

- Infant will reduce loss by lying in a flexed position and reducing blood flow to the skin (vasoconstriction)

FIGURE 7.12

As the first goal of an incubator is to ensure baby is within a thermal neutral environment, meaning baby neither loses heat or gains heat from its environment. There are of course exceptions to this rule, when baby is unable to maintain their own core body temperature due to size or illness, an additional heat is required. In the next four figures we will examine how heat is both lost and gained for a baby within an incubator.

Convection heat loss is firstly very much about the difference in temperature between the incubator air and baby's skin temperature (Figure 7.12). The air flow through an incubator must be sufficient to ensure a constant temperature throughout the canopy, but not that high that produces a 'wind chill' effect upon the baby's skin. As has been said before, an indication of baby becoming too cold is that baby raises their legs and pulls their arms closer to their bodies in order to minimize the amount of skin exposure to the air flow. If baby becomes too warm, perhaps due to a fever, then baby might spread their arms and legs in order to lose heat by convection to the air flowing over them.

By clothing, the baby heat loss to convection can be greatly reduced. A small cap, bonnet or light gown are often placed upon baby for this purpose. Sometimes when it is deemed possible, mothers are supported in dressing their babies in order help bonding between the two.

Incubator Heat loss - Evaporation

- A baby loses heat when water evaporates from the babies skin and breath

- Each 1ml. of water lost by evaporation removes 576 calories of heat. (Ref: Rutter)

- However, when a mature baby is heat stressed, active sweating occurs in an attempt to remove heat

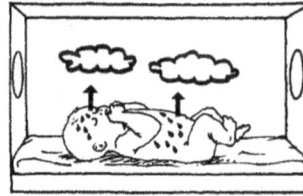

FIGURE 7.13

Incubators draw the air they need from the room's environment and in doing so that air will contain a certain percent of water vapor. The RH (relative humidity) at the room's air temperature of 25°C is approximately 30%RH, then once this air has been warmed to 36°C, the RH will drop to ~15%RH. For a baby with a well-developed skin, it is not seen as a problem and active humidification is not required.

As incubators are often used for the care of low birth weight and/or premature babies, it needs to be understood that these babies particularly will not at the time of birth have fully developed skin layers. Fully developed skin layers in human beings act to prevent water loss and retain heat. In a warm dry environment such as an incubator, baby's skin allows for the easy passage of water from their tissues through the skin as sweat and is then removed to the incubator air by evaporation (Figure 7.13). This process is detrimental to the baby as the loss of water represents a loss of much needed calories that if retained would produce the energy needed to maintain development, growth and fight any infections.

To counter this water vapor loss, it is possible to add humidification within the incubator which will reduce the water loss through the skin. This was previously done by using a sterile water chamber that was situated near the heater beneath the mattress. This would prove problematic as the sides of the Pperspex canopy would mist-up and impair visibility of the baby. A second draw back to this method was that in the warm damp environment of the incubator canopy bacteria would flourish from such sources as baby's faces and urine and present an infection hazard to the baby. Today, active humidification is provided by using a small Perspex head box that is placed over the head of the baby and supplied with humidified oxygen. One requirement when this is used is that an oxygen monitor sensor is also placed within the head box to ensure that the correct level of oxygen is administered. Excessive and prolonged over-oxygenation can have serious long-term neurological and sight damage issues for the baby.

Incubator Heat loss - Evaporation

- Skin water loss in infants below 28 weeks gestation, nursed naked in unhumidified incubators, compared with term infants.

- Losses are very high in the immediate neonatal period but then fall towards normal full term levels by two weeks of age

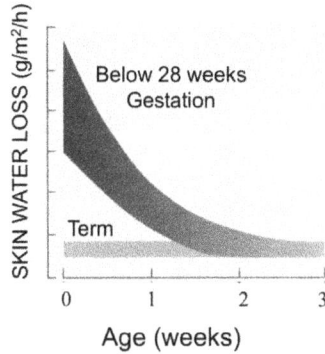

FIGURE 7.14

Figure 7.14 demonstrates just how much water loss a premature baby will suffer if there is no active humidification using a head box humidifier. The loss of water is conveyed as $g/m^2/h$ (g/m^2 of baby's skin surface/hour). As can be seen, the losses of a full-term baby are both low and constant, but the losses for a premature baby start some three to four times higher but reduce to the same level as a full-term baby only after 2–3 weeks. This reduction occurs as the baby's skin fully develops.

Incubator Heat loss - Evaporation

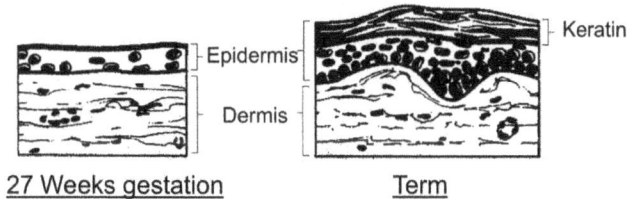

27 Weeks gestation   Term

- The skin of an infant at 27 weeks gestation compared with that of a term infant. The epidermis is thin, with little keratin formation

FIGURE 7.15

The most notable difference between a full-term baby's skin and a premature baby's is the development of the outer layer of keratin (Figure 7.15). This is a protein that is found not only in the outer skin layer but also in nails and hair. One of the most important properties it has is to retain water molecules with the epidermis and thus reduce baby's water loss. You might also notice that there is also an increase in the depth of the epidermis and the number of cells within it.

Incubator Heat loss - Radiation

- Loss of heat by radiating energy from the babies skin surface to all surrounding surfaces such as the walls of the room, windows and walls of the incubator

- It is independent of air temperature

- High levels of heat loss if the incubator is near a cold window

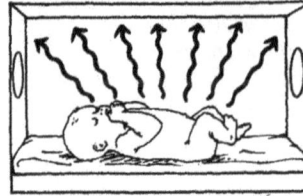

FIGURE 7.16

Humans, being warm-blooded, will generate thermal radiation that is emitted to the cooler walls, windows and doors. Because we are able to generate so much heat, this thermal radiation heat loss is negligible and has no impact on our well-being. Thermal radiation is where heat is transmitted from one object, in this case a baby, as infrared energy, to a cooler object such as a window surface. It will pass through the air in-between baby and the cooler window or wall without loss and can in certain circumstance prove to be a significant heat/energy drain on the baby. In the opposite direction if an incubator is placed close to a window, then sunlight can also pass infrared light through the window and into the incubator and then fall upon the baby which can produce excessive heat on the baby's skin (Figure 7.16). Although the incubator which may be in air mode appears to be controlling the incubator air temperature correctly, baby may begin to suffer heat stress from the sun's radiated heat. Care should always be taken when positioning an incubator to avoid direct sunlight and it should be kept at a distance from cool walls and windows. Incubators that are not being used should also be stored away from windows with direct sunlight as these incubators may act as greenhouses even if they are unplugged and not operating.

Many of today's incubators have an inner Perspex canopy wall lining that will significantly reduce baby's radiated heat loss as the infrared energy from baby is absorbed on this inner surface that will be at the same temperature of the incubator air flow.

Incubator Heat loss - Conduction

- Involves loss due to direct contact between baby's skin and cooler solid object such as the mattress

- This accounts for only a small fraction of the total heat loss for the infant. They may even gain heat by conduction.

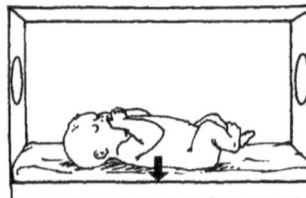

FIGURE 7.17

Conduction heat loss happens generally only when baby is first placed upon the mattress (Figure 7.17). If the incubator has been prepared, by running it in air mode at a temperature that will be close to the required temperature for baby, then the amount of conduction heat loss from baby will be minimal. If the baby is naked then the mattress may even act to insulate baby from heat losses.

- Many neonates are born with some degree of Jaundice. This appears as a yellowing of the skin and white's of their eyes.

- Jaundice is due to raised levels of Bilirubin which is a protein like secretion from the metabolism of red blood cells that is then broken down in the liver to be soluble and then passed from the body in urine and faeces.

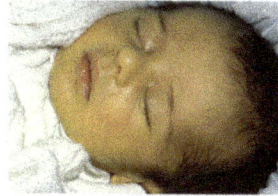

FIGURE 7.18 Pathological jaundice.

A large number of babies, both full term and premature are born with livers that are not fully developed. These babies appear to have a yellow skin color and yellow eyes (jaundice, Figure 7.18). Jaundice, if left untreated, can have serious health consequences, including nerve damage and in some cases death. There are also other reasons such as the breakdown of red blood cells due compatibility issues between the mother's blood type and baby's blood type. Jaundice is a condition where there is excessive bilirubin, red blood cells that have broken-down and are not processed in sufficient quantities in the liver and still circulate in the baby's blood stream. Normally these red blood cells will be processed in a well-developed liver and are passed out in feces and urine.

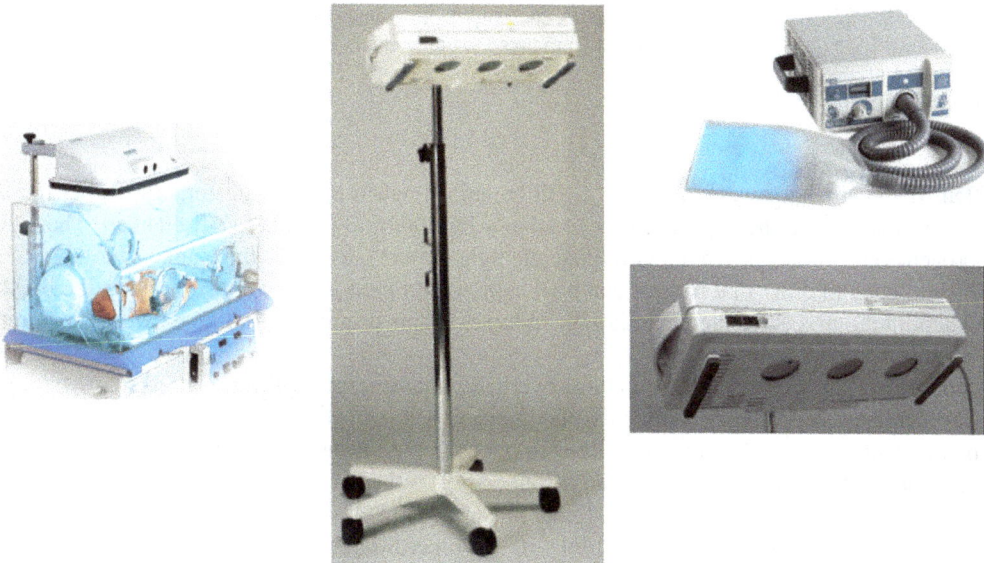

FIGURE 7.19 Neonatal phototherapy.

For many decades, clinicians such as midwifes and doctors understood that if baby was exposed to sunlight, then the jaundice appearance would improve over the following days. As direct sunlight is not always available, an alternative method was developed using artificial light. As white/visible light contains many wavelengths of light, from purple to deep red, research was undertaken to identify which wavelengths were most effective at bringing about the metabolizing (breaking down) of these discarded red blood cells in order to assist the liver in their removal. It was not long before that blue/green light between 460 and 490 nanometers (nm) was identified as the most effective. Light bulbs and lamps that produced these wavelengths were then produced and incorporated in phototherapy devices and the concept of neonatal phototherapy was adopted (Figure 7.19).

Just why light, and particularly light in the blue, blue-green spectrum (400–550 nm wavelength), is able to enhance the breakdown of bilirubin is still not fully understood. It often appears within the first 24 hours after birth and is in the majority of cases harmless. Jaundice does understandably cause some concern in parents and this anxiety can be transmitted to the infant. For this reason and clinical reasons, phototherapy is often an appropriate therapy for the treatment of the condition (Figure 7.20).

Phototherapy Treatment

- Considerations effecting Phototherapy use are the following:
  - As much as possible of the infants surface area should be available for radiation with the light.
  - Great care must be taken to cover the infants eyes during therapy.
  - Heat generated by the Fluorescent/Quart halogen/Gas discharge/Light Emitting diodes must not be radiated to the infant.

FIGURE 7.20

Often when baby has severe jaundice, it will be nursed naked in an incubator, with either the use of an overhead phototherapy unit, or lain upon a light mattress/blanket (bili blanket). These two approaches ensure the maximum of therapeutic light possible. Recently the blankets have been further developed in order that parents can hold baby within their arms while still within the blanket and continuing to receive the phototherapy treatment. This promotes bonding between the parents and the baby.

For a baby nursed beneath a standard overhead phototherapy unit, great care must be taken to protect baby's eyes. This is done by placing eye pads over their eyes. The pads are regularly replaced and the eyes cleaned. One other important issue in using any phototherapy treatment is that the phototherapy device does not produce heat either directly to the baby's skin or to the surfaces and air within the incubator. Today most phototherapy units are constructed using light emitting diodes (LEDs) which, compared with the older fluorescent/quart halogen/gas discharge lamps, produce only a fraction of the heat the older light sources produced. LEDs also have the advantage of superior constant light output levels for many thousands of hours over the older light sources.

Phototherapy Treatment

- Considerations effecting Phototherapy use are the following:
  - Height of the light source above the infant

  - Often a white light source is included along with the therapeutic blue or blue/green light in order that the true colour of the infants skin can be assessed. This also helps prevent headaches and nausea in staff and parents.

FIGURE 7.21

As with any form of light, the more distant the light source is from the baby, the less therapeutic effect the light will have. This is particularly the case with an overhead phototherapy unit (Figure 7.21). If the unit is placed on or very close to the incubator canopy, then it is possible for the phototherapy unit to produce unwelcome heat in the Perspex incubator canopy. Allowing a sufficient distance between the phototherapy unit and the incubator canopy allows the room's ventilation to remove this unwelcome heat source while still allowing enough therapeutic light for the baby.

When a baby is placed under the blue/green light, it becomes more difficult to accurately assess baby's true skin color. For this reason, a standard white/visible light may also be incorporated into the phototherapy unit in order to allow for accurate skin color assessment. Also, the use of blue/green light alone has been known to cause parents and staff after a time to feel unwell with headaches and nausea.

Phototherapy
Light output
meter

Bili Check
Non-invasive
Bilirubin
Assessment Tool,

FIGURE 7.22   Phototherapy light output meter & Bilirubin diagnosis tools.

Figure 7.22 shows two devices commonly used in the neonatal unit. The first is the phototherapy light output meter used by technicians to measure the amount of light at the required blue/green wavelengths a phototherapy device produces at a given distance. Regular testing of the amount of light is very important as all types of light sources will eventually lessen their outputs and will require replacement.

The second device shown is a bilirubin assessment tool, transcutaneous bilirubinometer that is used to measure the amount of jaundice a baby has by using a light source that is placed against baby's forehead or chest It then transmits visible light and the amount of light refracted back on to an inbuilt light detector will provide a measurement number indicating the level of bilirubin in the baby's blood stream. This non-invasive system is an alternative to a heel prick blood sample but is not as accurate.

For Product Safety Concerns and Information please contact our EU
representative GPSR@taylorandfrancis.com
Taylor & Francis Verlag GmbH, Kaufingerstraße 24, 80331 München, Germany